Industrial

Discipline-Specific Review for the FE/EIT Exam

Second Edition

University of Missouri, Columbia Department of Industrial and Manufacturing Systems Engineering with Michael R. Lindeburg, PE

PPI®

PPI2PASS.COM

A **KAPLAN** COMPANY

Register Your Book at ppi2pass.com/register

- Receive the latest exam news.
- Obtain exclusive exam tips and strategies.
- Receive special discounts.

Report Errors for This Book

PPI is grateful to every reader who notifies us of a possible error. Your feedback allows us to improve the quality and accuracy of our products. Report errata at **ppi2pass.com/errata**.

INDUSTRIAL DISCIPLINE-SPECIFIC REVIEW FOR THE FE/EIT EXAM
Second Edition

Current printing of this edition: 6

Printing History

date	edition number	printing number	update
Apr 2016	2	4	Minor cover updates.
Feb 2018	2	5	Minor corrections. Minor cover updates.
Jun 2019	2	6	Minor corrections. Minor cover updates.

Printed in the United States of America.

PPI
1250 Fifth Avenue
Belmont, CA 94002
(650) 593-9119
ppi2pass.com

ISBN: 978-1-59126-068-4

Library of Congress Control Number: 2007932503

Table of Contents

Preface and Acknowledgments

This book is one in a series intended for engineers and students who are taking a discipline-specific (DS) afternoon session of the Fundamentals of Engineering (FE) exam.

The topics covered in the DS afternoon FE exams are completely different from the topics covered in the morning session of the FE exam. Since this book only covers one discipline-specific exam, it provides exam-level problems that are like those found on the afternoon half of the FE exam for the industrial discipline.

This book is intended to be a quick review of the material unique to the afternoon session of the industrial engineering exam. The material presented covers the subjects most likely to be on the exam. This book is not a thorough treatment of the exam topics. Its objective is to prepare you with enough knowledge to pass. As much as practical, this book uses the notation given in the NCEES Handbook.

This book consolidates 191 practical review problems, covering all of the discipline-specific exam topics. All problems include full solutions.

The problems in this book's first edition were developed by the University of Missouri–Columbia, Department of Industrial Engineering, and the problems for the second edition were developed by James S. Noble, PhD, PE, following the format, style, subject breakdown, and guidelines that I provided.

In developing this book, the NCEES Handbook and the breakdown of problem types published by NCEES were my guide for problem types and scope of coverage. However, as with most standardized tests, there is no guarantee that any specific problem type will be encountered. It is expected that minor variations in problem content will occur from exam to exam.

As with all of PPI's books, the problems in this book are original and have been ethically derived. Although examinee feedback was used to determine its content, this book contains problems that are only *like* those that are on the exam. There are no actual exam problems in this book.

This book was designed to complement my *FE Review Manual*, which you will also need to prepare for the FE exam. The *FE Review Manual* is PPI's most popular study guide for both the morning and afternoon general exams. It and the *Engineer-In-Training Reference Manual* have been the most popular review books for this exam for more than 25 years.

You cannot prepare adequately without your own copy of the NCEES Handbook. This document contains the data and formulas that you will need to solve both the general and the discipline-specific problems. A good way to become familiar with it is to look up the information, formulas, and data that you need while trying to work practice problems.

Exam-prep books are always works in progress. By necessity, a book will change as the exam changes. Even when the exam format doesn't change for a while, new problems and improved explanations can always be added. I encourage you to provide comments via PPI's errata reporting page, **www.ppi2pass.com/errata**. You will find all verified errata there. I appreciate all feedback.

Best of luck to you in your pursuit of licensure.

Michael R. Lindeburg, PE

Engineering Registration in the United States

ENGINEERING REGISTRATION

Engineering registration (also known as *engineering licensing*) in the United States is an examination process by which a state's board of engineering licensing (i.e., registration board) determines and certifies that you have achieved a minimum level of competence. This process protects the public by preventing unqualified individuals from offering engineering services.

Most engineers do not need to be registered. In particular, most engineers who work for companies that design and manufacture products are exempt from the licensing requirement. This is known as the *industrial exemption*. Nevertheless, there are many good reasons for registering. For example, you cannot offer consulting engineering design services in any state unless you are registered in that state. Even within a product-oriented corporation, however, you may find that registered engineers have more opportunities for employment and advancement.

Once you have met the registration requirements, you will be allowed to use the titles Professional Engineer (PE), Registered Engineer (RE), and Consulting Engineer (CE).

Although the registration process is similar in all 50 states, each state has its own registration law. Unless you offer consulting engineering services in more than one state, however, you will not need to register in other states.

The U.S. Registration Procedure

To become a registered engineer in the United States, you will need to pass two eight-hour written examinations. The first is the *Fundamentals of Engineering Examination*, also known as the *Engineer-In-Training Examination* and the *Intern Engineer Exam*. The initials FE, EIT, and IE are also used. This exam covers basic subjects from the mathematics, physics, chemistry, and engineering classes you took during your first four university years. In rare cases, you may be allowed to skip this first exam.

The second eight-hour exam is the *Principles and Practice of Engineering Exam*. The initials PE are also used. This exam is on topics within a specific discipline, and only covers subjects that fall within that area of specialty.

Most states have similar registration procedures. However, the details of registration qualifications, experience requirements, minimum education levels, fees, oral interviews, and exam schedules vary from state to state. For more information, contact your state's registration board (**www.ppi2pass.com/stateboards**).

National Council of Examiners for Engineering and Surveying

The National Council of Examiners for Engineering and Surveying (NCEES) in Clemson, South Carolina, produces, distributes, and scores the national FE and PE exams. The individual states purchase the exams from NCEES and administer them themselves. NCEES does not distribute applications to take the exams, administer the exams or appeals, or notify you of the results. These tasks are all performed by the states.

Reciprocity Among States

With minor exceptions, having a license from one state will not permit you to practice engineering in another state. You must have a professional engineering license from each state in which you work. For most engineers, this is not a problem, but for some, it is. Luckily, it is not too difficult to get a license from every state you work in once you have a license from one state.

All states use the NCEES exams. If you take and pass the FE or PE exam in one state, your certificate will be honored by all of the other states. Although there may be other special requirements imposed by a state, it will not be necessary to retake the FE and PE exams. The issuance of an engineering license based on another state's license is known as *reciprocity* or *comity*.

The simultaneous administration of identical exams in all states has led to the term *uniform examination*. However, each state is still free to choose its own minimum passing score and to add special questions and requirements to the examination process. Therefore, the use of a uniform exam has not, by itself, ensured reciprocity among states.

THE FE EXAM

Applying for the Exam

Each state charges different fees, specifies different requirements, and uses different forms to apply for the exam. Therefore, it will be necessary to request an application from the state in which you want to become registered. Generally, it is sufficient for you to phone for this application. You'll find contact information (websites, telephone numbers, email addresses, etc.) for all U.S. state and territorial boards of registration at **www.ppi2pass.com/stateboards**.

Keep a copy of your exam application, and send the original application by certified mail, requesting a delivery receipt. Keep your proof of mailing and delivery with your copy of the application.

Exam Dates

The national FE and PE exams are administered twice a year (usually in mid-April and late October), on the same weekends in all states. For a current exam schedule, check **www.ppi2pass.com/fefaq**.

FE Exam Format

The NCEES Fundamentals of Engineering examination has the following format and characteristics.

- There are two four-hour sessions separated by a one-hour lunch.

- Examination questions are distributed in a bound examination booklet. A different exam booklet is used for each of the two sessions.

- Formulas and tables of data needed to solve questions in the exams are found in either the NCEES Handbook or in the body of the question statement itself.

- The morning session (also known as the *A.M. session*) has 120 multiple-choice questions, each with four possible answers lettered (A) through (D). Responses must be recorded with a pencil provided by NCEES on special answer sheets. No credit is given for answers recorded in ink.

- Each problem in the morning session is worth one point. The total score possible in the morning is 120 points. Guessing is valid; no points are subtracted for incorrect answers.

- There are questions on the exam from most of the undergraduate engineering degree program subjects. Questions from the same subject are all grouped together, and the subjects are labeled. The percentages of questions for each subject in the morning session are given in the following table.

Morning FE Exam Subjects

subject	percentage of questions (%)
chemistry	9
computers	7
electricity and magnetism	9
engineering economics	8
engineering probability and statistics	7
engineering mechanics (statics and dynamics)	10
ethics and business practices	7
fluid mechanics	7
material properties	7
mathematics	15
strength of materials	7
thermodynamics	7

- There are seven different versions of the afternoon session (also known as the *P.M. session*), six of which correspond to specific engineering disciplines: chemical, civil, electrical, environmental, industrial, and mechanical engineering.

The seventh version of the afternoon exam is a general examination suitable for anyone, but in particular, for engineers whose specialties are not one of the other six disciplines. Though the subjects in the general afternoon exam are similar to the morning subjects, the questions are more complex—hence their double weighting. Questions on the afternoon exam are intended to cover concepts learned in the last two years of a four-year degree program. Unlike morning questions, these questions may deal with more than one basic concept per question.

Each version of the afternoon session consists of 60 questions. All questions are mandatory. Questions in each subject may be grouped into related problem sets containing between two and ten questions each.

The percentages of questions for each subject in the general afternoon session exam are given in the following table.

Afternoon FE Exam Subjects (General Exam)

subject	percentage of questions (%)
advanced engineering mathematics	10
application of engineering mechanics	13
biology	5
electricity and magnetism	12
engineering economics	10
engineering probability and statistics	9
engineering of materials	11
fluids	15
thermodynamics and heat transfer	15

Each of the discipline-specific afternoon examinations covers a substantially different body of knowledge than the morning exam. The percentages of questions for each subject in the industrial discipline-specific afternoon session exam are as follows.

Afternoon FE Exam Subjects
(Industrial DS Exam)

subject	percentage of questions (%)
Engineering Economics	15
Probability and Statistics	15
Modeling and Computation	12
Industrial Management	10
Manufacturing and Production Systems	13
Facilities and Logistics	12
Human Factors, Productivity, Ergonomics, and Work Design	12
Quality	11

Some afternoon questions stand alone, while others are grouped together, with a single problem statement that describes a situation followed by two or more questions about that situation. All questions are multiple-choice. You must choose the best answer from among four, lettered (A) through (D).

- Each question in the afternoon is worth two points, making the total possible score 120 points.

- The scores from the morning and afternoon sessions are added together to determine your total score. No points are subtracted for guessing or incorrect answers. Both sessions are given equal weight. It is not necessary to achieve any minimum score on either the morning or afternoon sessions.

- All grading is done by computer optical sensing.

Use of SI Units on the FE Exam

Metric questions are used in all subjects, except some civil engineering and surveying subjects that typically use only customary U.S. (i.e., English) units. SI units are consistent with ANSI/IEEE standard 268 (the American Standard for Metric Practice). Non-SI metric units might still be used when common or where needed for consistency with tabulated data (e.g., use of bars in pressure measurement).

Grading and Scoring the FE Exam

The FE exam is not graded on the curve, and there is no guarantee that a certain percentage of examinees will pass. Rather, NCEES uses a modification of the Angoff procedure to determine the suggested passing score (the cutoff point or cut score).

With this method, a group of engineering professors and other experts estimate the fraction of minimally qualified engineers who will be able to answer each question

correctly. The summation of the estimated fractions for all test questions becomes the passing score. Because the law in most states requires engineers to achieve a score of 70% to become licensed, you may be reported as having achieved a score of 70% if your raw score is greater than the passing score established by NCEES, regardless of the raw percentage. The actual score may be slightly more or slightly less than 110 as determined from the performance of all examinees on the equating subtest.

About 20% of the FE exam questions are repeated from previous exams—this is the *equating subtest*. Since the scores of previous examinees on the equating subtest are known, comparisons can be made between the two exams and examinee populations. These comparisons are used to adjust the passing score.

The individual states are free to adopt their own passing score, but all adopt NCEES's suggested passing score because the states believe this cutoff score can be defended if challenged.

You will receive the results within 12 weeks of taking the exam. If you pass, you will receive a letter stating that you have passed. If you fail, you will be notified that you failed and be provided with a diagnostic report.

Permitted Reference Material

Since October 1993, the FE exam has been what NCEES calls a "limited-reference" exam. This means that no books or references other than those supplied by NCEES may be used. Therefore, the FE exam is really an "NCEES-publication only" exam. NCEES provides its own Supplied-Reference Handbook for use during the examination. No books from other publishers may be used.

CALCULATORS

In most states, battery- and solar-powered, silent calculators can be used during the exam, although printers cannot be used. (Solar-powered calculators are preferable because they do not have batteries that run down.) In most states, programmable, preprogrammed, and business/finance calculators are allowed. Similarly, nomographs and specialty slide rules are permitted. To prevent unauthorized transcription and redistribution of the exam questions, calculators with communication or text-editing capabilities are banned from all NCEES exam sites. You cannot share calculators with other examinees. For a list of allowed calculators check **www.ppi2pass.com/calculators**.

It is essential that a calculator used for engineering examinations have the following functions.

- trigonometric functions
- inverse trigonometric functions
- hyperbolic functions
- pi
- square root and x^2
- common and natural logarithms
- y^x and e^x

For maximum speed, your calculator should also have or be programmed for the following functions.

- extracting roots of quadratic and higher-order equations
- converting between polar (phasor) and rectangular vectors
- finding standard deviations and variances
- calculating determinants of 3×3 matrices
- linear regression
- economic analysis and other financial functions

STRATEGIES FOR PASSING THE FE EXAM

The most successful strategy for passing the FE exam is to prepare in all of the exam subjects. Do not limit the number of subjects you study in hopes of finding enough questions in your strongest areas of knowledge to pass.

Fast recall and stamina are essential to doing well. You must be able to quickly recall solution procedures, formulas, and important data. You will not have time during the exam to derive solutions methods—you must know them instinctively. This ability must be maintained for eight hours. Be sure to gain familiarity with the NCEES Handbook by using it as your only reference for some of the problems you work during study sessions.

In order to get exposure to all exam subjects, it is imperative that you develop and adhere to a review schedule. If you are not taking a classroom review course (where the order of your preparation is determined by the lectures), prepare your own review schedule.

There are also physical demands on your body during the exam. It is very difficult to remain alert and attentive for eight hours or more. Unfortunately, the more time you study, the less time you have to maintain your physical condition. Thus, most examinees arrive at the exam site in peak mental condition but in deteriorated physical condition. While preparing for the FE exam is not the only good reason for embarking on a physical conditioning program, it can serve as a good incentive to get in shape.

It will be helpful to make a few simple decisions prior to starting your review. You should be aware of the different options available to you. For example, you should decide early on to

- use SI units in your preparation
- perform electrical calculations with effective (rms) or maximum values
- take calculations out to a maximum of four significant digits
- prepare in all exam subjects, not just your specialty areas

At the beginning of your review program, you should locate a spare calculator. It is not necessary to buy a spare if you can arrange to borrow one from a friend or the office. However, if possible, your primary and spare calculators should be identical. If your spare calculator is not identical to the primary calculator, spend some time familiarizing yourself with its functions.

A Few Days Before the Exam

There are a few things you should do a week or so before the exam date. For example, visit the exam site in order to find the building, parking areas, examination room, and rest rooms. You should also make arrangements now for child care and transportation. Since the exam does not always start or end at the designated times, make sure that your child care and transportation arrangements can tolerate a late completion.

Next in importance to your scholastic preparation is the preparation of your two examination kits. The first kit consists of a bag or box containing items to bring with you into the examination room.

[ ]   letter admitting you to the exam
[ ]   photographic identification
[ ]   main calculator
[ ]   spare calculator
[ ]   extra calculator batteries
[ ]   unobtrusive snacks
[ ]   travel pack of tissues
[ ]   headache remedy
[ ] $2.00 in change
[ ]   light, comfortable sweater
[ ]   loose shoes or slippers
[ ]   handkerchief
[ ]   cushion for your chair
[ ]   small hand towel
[ ]   earplugs
[ ]   wristwatch with alarm
[ ]   wire coat hanger
[ ]   extra set of car keys

The second kit consists of the following items and should be left in a separate bag or box in your car in case you need them.

[ ]   copy of your application
[ ]   proof of delivery
[ ]   this book
[ ]   other references
[ ]   regular dictionary
[ ]   scientific dictionary
[ ]   course notes in three-ring binders
[ ]   instruction booklets for all your calculators
[ ]   light lunch
[ ]   beverages in thermos and cans
[ ]   sunglasses
[ ]   extra pair of prescription glasses
[ ]   raincoat, boots, gloves, hat, and umbrella
[ ]   street map of the exam site
[ ]   note to the parking patrol for your windshield explaining where you are, what you are doing, and why your time may have expired
[ ]   battery-powered desk lamp

The Day Before the Exam

Take the day before the exam off from work to relax. Do not cram the last night. A good prior night's sleep is the best way to start the exam. If you live far from the exam site, consider getting a hotel room in which to spend the night.

Make sure your exam kits are packed and ready to go.

The Day of the Exam

You should arrive at least 30 minutes before the exam starts. This will allow time for finding a convenient parking place, bringing your materials to the exam room, and making room and seating changes. Be prepared, though, to find that the examination room is not open or ready at the designated time.

Once the examination has started, consider the following suggestions.

- Set your wristwatch alarm for five minutes before the end of each four-hour session, and use that remaining time to guess at all of the remaining unsolved problems. Do not work up until the very end. You will be successful with about 25% of your guesses, and these points will more than make up for the few points you might earn by working during the last five minutes.

- Do not spend more than two minutes per morning question. (The average time available per problem is two minutes.) If you have not finished a question in that time, make a note of it and move on.

- Do not ask your proctors technical questions. Even if they are knowledgeable in engineering, they will not be permitted to answer your questions.

- Make a quick mental note about any problems for which you cannot find a correct response or for which you believe there are two correct answers. Errors in the exam are rare, but they do occur. Being able to point out an error later might give you the margin you need to pass. Since such problems are almost always discovered during the scoring process and discounted from the exam, it is not necessary to tell your proctor, but be sure to mark the one best answer before moving on.

- Make sure all of your responses on the answer sheet are dark and completely fill the bubbles.

Common Questions About the DS Exam

Q: Do I have to take the DS exam?

A: Most people do not have to take the DS exam and may elect the general exam option. The state boards do not care which afternoon option you choose, nor do employers. In some cases, an examinee still in an undergraduate degree program may be required by his or her university to take a specific DS exam.

Q: Do all mechanical, civil, electrical, chemical, industrial, and environmental engineers take the DS exam?

A: Originally, the concept was that examinees from the "big five" disciplines would take the DS exam, and the general exam would be for everyone else. This remains just a concept, however. A majority of engineers in all of the disciplines apparently take the general exam.

Q: When do I elect to take the DS exam?

A: You will make your decision on the afternoon of the FE exam, when the exam booklet (containing all of the DS exams) is distributed to you.

Q: Where on the application for the FE exam do I choose which DS exam I want to take?

A: You don't specify the DS option at the time of your application.

Q: After starting to work on either the DS or general exam, can I change my mind and switch options?

A: Yes. Theoretically, if you haven't spent too much time on one exam, you can change your mind and start a different one. (You might need to obtain a new answer sheet from the proctor.)

Q: After I take the DS exam, does anyone know that I took it?

A: After you take the FE exam, only NCEES and your state board will know whether you took the DS or general exam. Such information may or may not be retained by your state board.

Q: Will my DS FE certificate be recognized by other states?

A: Yes. All states recognize passing the FE exam and do not distinguish between the DS and general afternoon portions of the FE exam.

Q: Is the DS FE certificate "better" than the general FE certificate?

A: There is no difference. No one will know which option you chose. It's not stated on the certificate you receive from your state.

Q: What is the format of the DS exam?

A: The DS exam is 4 hours long. There are 60 problems, each worth 2 points. The average time per problem is 4 minutes. Each problem is multiple choice with 4 answer choices. Most problems require the application of more than one concept (i.e., formula).

Q: Is there anything special about the way the DS exam is administered?

A: In all ways, the DS and general afternoon exam are equivalent. There is no penalty for guessing. No credit is given for scratch pad work, methods, etc.

Q: Are the answer choices close or tricky?

A: Answer choices are not particularly close together in value, so the number of significant digits is not going to be an issue. Wrong answers, referred to as "distractors" by NCEES, are credible. However, the exam is not "tricky"; it does not try to mislead you.

Q: Are any problems in the afternoon session related to each other?

A: Several questions may refer to the same situation or figure. However, NCEES has tried to make all of the questions independent. If you make a mistake on one question, it shouldn't carry over to another.

Q: Is there any minimum passing score for the DS exam?

A: No. It is the total score from your morning and afternoon sessions that determines your passing, not the individual session scores. You do not have to "pass" each session individually.

Q: Is the general portion easier, harder, or the same as the DS exams?

A: Theoretically, all of the afternoon options are the same. At least, that is the intent of offering the specific options—to reduce the variability. Individual passing rates, however, may still vary 5% to 10% from exam to exam. (PPI lists the most recent passing statistics for the various discipline-specific options on its website, **www.ppi2pass.com/fepassrates**.)

Q: Do the DS exams cover material at the undergraduate or graduate level?

A: Like the general exam, test topics come entirely from the typical undergraduate degree program. However, the emphasis is primarily on material from the third and fourth year of your program. This may put examinees who take the exam in their junior year at a disadvantage.

Q: Do you need practical work experience to take the DS exam?

A: No.

Q: Does the DS exam also draw on subjects that are in the general exam?

A: Yes. The dividing line between general and DS topics is often indistinct.

Q: Is the DS exam in customary U.S. or SI units?

A: The DS exam is nearly entirely in SI units. A few exceptions exist for some engineering subjects (surveying, hydrology, code-based design, etc.) where current common practice uses only customary U.S. units.

Q: Does the NCEES Handbook cover everything that is on the DS exam?

A: No. You may be tested on subjects that are not in the NCEES Handbook. However, NCEES has apparently adopted an unofficial policy of providing any necessary information, data, and formulas in the stem of the question. You will not be required to memorize any formulas.

Q: Is everything in the DS portion of the NCEES Handbook going to be on the exam?

A: Apparently, there is a fair amount of reference material that isn't needed for every exam. There is no way, however, to know in advance what material is needed.

Q: How long does it take to prepare for the DS exam?

A: Preparing for the DS exam is similar to preparing for a mini PE exam. Engineers typically take two to four months to complete a thorough review for the PE exam. However, examinees who are still in their degree program at a university probably aren't going to spend more than two weeks thinking about, worrying about, or preparing for the DS exam. They rely on their recent familiarity with the subject matter.

Q: If I take the DS exam and fail, do I have to take the DS exam the next time?

A: No. The examination process has no memory.

Q: Where can I get even more information about the DS exam?

A: If you have internet access, visit the Exam FAQs and the Engineering Exam Forum at PPI's website (**www.ppi2pass.com/fe**).

How to Use This Book

HOW EXAMINEES CAN USE THIS BOOK

This book is divided into three parts: The first part consists of 71 representative practice problems covering all of the topics in the afternoon DS exam. Sixty problems corresponds to the number of problems in the afternoon DS exam. You may time yourself by allowing approximately 4 minutes per problem when attempting to solve these problems, but that was not my intent when designing this book. Since the solution follows directly after each problem in this section, I intended for you to read through the problems, attempt to solve them on your own, become familiar with the support material in the official NCEES Handbook, and accumulate the reference materials you think you will need for additional study.

The second and third parts of this book consists of two complete sample examinations that you can use as sources of additional practice problems or as timed diagnostic tools. They also contain 60 problems, and the number of problems in each subject corresponds to the breakdown of subjects published by NCEES. Since the solutions to these parts of the book are consolidated at the end, it was my intent that you would solve these problems in a realistic mock-exam mode.

You should use the NCEES Handbook as your only reference during the mock exams.

The morning general exam and the afternoon DS exam essentially cover two different bodies of knowledge. It takes a lot of discipline to prepare for two standardized exams simultaneously. Because of that (and because of my good understanding of human nature), I suspect that you will be tempted to start preparing for your chosen DS exam only after you have become comfortable with the general subjects. That's actually quite logical, because if you run out of time, you will still have the general afternoon exam as a viable option.

If, however, you are limited in time to only two or three months of study, it will be quite difficult to do a thorough DS review if you wait until after you have finished your general review. With a limited amount of time, you really need to prepare for both exams in parallel.

HOW INSTRUCTORS CAN USE THIS BOOK

The availability of the discipline-specific FE exam has greatly complicated the lives of review course instructors and coordinators. The general consensus is that it is essentially impossible to do justice to all of the general FE exam topics and then present a credible review for each of the DS topics. Increases in course cost, expenses, course length, and instructor pools (among many other issues) all conspire to create quite a difficult situation.

One-day reviews for each DS subject are subject-overload from a reviewing examinee's standpoint. Efforts to shuffle FE students over the parallel PE review courses meet with scheduling conflicts. Another idea, that of lengthening lectures and providing more in-depth coverage of existing topics (e.g., covering transistors during the electricity lecture), is perceived as a misuse of time by a majority of the review course attendees. Is it any wonder that virtually every FE review course in the country has elected to only present reviews for the general afternoon exam?

But, while more than half of the examinees elect to take the general afternoon exam, some may actually be required to take a DS exam. This is particularly the case in some university environments where the FE exam has become useful as an "outcome assessment tool." Thus, some method of review is still needed.

Since most examinees begin reviewing approximately two to three months before the exam (which corresponds to when most review courses begin), it is impractical to wait until the end of the general review to start the DS review. The DS review must proceed in parallel with the general review.

In the absence of parallel DS lectures (something that isn't yet occurring in too many review courses), you may want to structure your review course to provide lectures only on the general subjects. Your DS review could be assigned as "independent study," using chapters and problems from this book. Thus, your DS review would consist of distributing this book with a schedule of assignments. Your instructional staff could still provide assistance on specific DS problems, and completed DS assignments could still be recorded.

The final chapter on incorporating DS subjects into review courses has yet to be written. Like the landscape architect who waits until a well-worn path appears through the plants before placing stepping stones, we need to see how review courses do it before we can give any advice.

Nomenclature

ENGINEERING ECONOMICS

A	uniform amount per period	$
ATCF	after-tax cash flow	$
B	benefits	–
BV_j	book value at end of year j	$
C	cost	$
C_F	fixed cost	$
C_v	variable cost	$
D	production quantity	–
D_j	depreciation in year j	$
i	interest rate	%
I	construction cost index	–
I	initial investment	$
IP	interest payment	$
F	future worth, value, or amount	$
G	gradient amount	$
m	number of compounding periods per year	–
n	expected life of an asset	yr
n	number of compounding periods	–
p	selling price	$
P	present worth, value, or amount	$
PBP	pay-back period	yr
PP	principle payment	$
r	nominal annual interest rate per year	%
R	revenue	$
S_n	salvage value in year n	$
t	tax rate	%
T	project duration, time-in-system	yr

PROBABILITY AND STATISTICS

$\overline{d}$	average difference between two systems	–
d_i	difference in pair i	–
MSE	mean square error	–
MST	mean square error between treatments	–
n	number of replications	–
n	sample size	–
N	number of observations	–
P	probability	–
s	sample standard deviation	–
SS	sum of squares	–
SS_{error}	error sum of squares	–

SS_{total}	total sum-of-squares errors	–
t	total length	ft
T	time-in-system	min
T	total observations	–
$\overline{x}$	sample mean	–
z	standard normal variant	–
λ	average arrival rate	min^{-1}
μ	true mean	–
π	steady-state probability	–
σ	standard deviation	–

MODELING AND COMPUTATION

∇f	gradient of the function	–
L	expected number of units	–
n	number of units in a system	–
P	probability	%
W	waiting time	min
Y	output	–
Λ	transition rate matrix	–
λ	mean arrival rate	min^{-1}
λ_{ij}	transition rate from state i to state j	–
μ	mean service rate (constant)	min^{-1}
π	steady-state probability vector	–
ρ	server utilization	–

INDUSTRIAL MANAGEMENT

A_{ij}	optimistic duration for activity (i,j)	hr
B_{ij}	most likely duration for activity (i,j)	hr
C_{ij}	pessimistic duration for activity (i,j)	hr
μ_{ij}	mean duration for activity (i,j)	hr
σ_{ij}^2	variance of duration of activity (i,j)	hr

MANUFACTURING AND PRODUCTION SYSTEMS

A	cross-sectional area	in^2
A	set up cost per lot	$
C	capacity	–
C	purchase price per unit	$
C_{ij}	production time for product i on machine j	hr
$\hat{d}_t$	forecasted demand for time t	–
d_t	demand for time t	–
D	demand	–

E	actual performance of machine	–
f	feed rate	in/min
F	forecast	–
F	total number of machines ceiling	–
F_t	number of employees fired at beginning of period t	–
h	holding cost per unit	$
H	available units per time period	–
H	productive time per person	hr
H_t	employees hired at beginning of period t	–
K	cost	$
L	lead time	hr
L	length of cut	in
M	number of machines	–
n	number of observations, products, or units	–
P	set up time	hr
$P_{i,k}$	scrap rate for machine k part i	–
Q^*	economic production quantity	–
r	radius	in
r	cutting rate	in/min
r	interest rate on inventory	%/yr
R	annual replenishment rate	–/yr
R	machine availability	–
s	spindle speed	in/sec, ft/min
S	standard time	hr
S_w	shifts per week	–
t	time	hr
T	production time	hr
V	volume	–
W	workers per shift	–
W_t	employees needed in period t	–
α	safety factor	–
α	smoothing constant	–
λ	average demand per period	–/hr

FACILITIES AND LOGISTICS

C	time in production	hr
D	distance	ft
F	flow	–
H	productive time	hr
L	lead time	hr
L	mean arrival rate	–/hr
M	number of machines	–
N	number of products	–
P	production rate	–/hr
T	production time	hr
TC	material handling cost	$
w	weighting factor	–
W	minimum number of workers per shift	–
S	number of servers	–

S	number of shifts	–
α	safety factor	–
μ	mean service time	hr
ρ	utilization factor	–

HUMAN FACTORS, PRODUCTIVITY, ERGONOMICS, AND WORK DESIGN

A	allowances	%
AT	allowed time	hr
D	absolute error	–
D	vertical lifted distance	in
F	average lifts per minute	–/min
F	frequency	various
$F_{(max)}$	maximum frequency of lifting	–/min
h	thickness of support	in
H	horizontal distance from hands to body's center	in
K	occurrences per cycle	–
K	time to produce the first output unit	hr
MPL	maximum permissible load	lbf
n	number of elements, observations	–
n	required sample size	–
OT	observed time	hr
p	proportion activity observed or occurring	–
R	objective rating	%
s	element standard deviation	various
s	learning curve slope parameter	–
S	degree of accuracy required	%
ST	standard time	min
T	thickness	in
T	thigh clearance height	in
u	output unit number	–
V	vertical distance from hands to floor	in
VCU	variable cost per unit	$/unit
x	eye height from floor	in
x	number of times the activity is observed	–
z	standard normal distribution variant	–
Z	output production time	hr
ε	desired accuracy	%

QUALITY

c	acceptance number	–
C_{pk}	process capability index	–
C_x	process capability measure	various
d	number of defects index	–
d_2	constant factor for sample size n	–
D_3	constant factor for LCL	–
D_4	constant factor for UCL	–

K	trial control limits, number of sample ranges	–
LCL	lower control unit	various
LSL	lower specification limit	various
n	random sample size	–
N	lot size	–
p	fraction of defective items	–
R	range	various
UCL	upper control limit	various
USL	upper specification limit	various
X	individual observation	various
z	standard normal distribution variant	various
μ	population mean	various
σ	standard deviation	various
$\hat{\sigma}_X$	population standard deviation	various

Practice Problems

ENGINEERING ECONOMICS

Problem 1

Kelly just won the state lottery. The $2,000,000 jackpot will be paid in 20 annual installments of $100,000. The first installment is given to Kelly immediately. The interest rate is 6% compounded yearly. The present worth of Kelly's lottery winnings (at the time of receiving the first installment) is most nearly

    (A) $1,120,000
    (B) $1,150,000
    (C) $1,220,000
    (D) $1,900,000

Solution

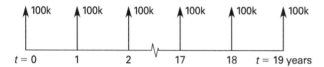

Use the uniform series present worth factor to find the present worth of the last 19 installments at the time the first installment is made. Add this value to the first installment to determine the total present worth of Kelly's lottery winnings.

$$A = \text{uniform series of end of}$$
$$\text{compounding period cash flows}$$
$$= \$100,000$$

$$i = \text{annual interest rate} = 6\%$$

$$n = \text{number of compounding periods} = 19 \text{ years}$$

$$P = \text{present equivalent value of a cash flow}$$
$$\text{or series of cash flows}$$
$$= \$100,000 + (\$100,000)(P/A, 6\%, 19)$$
$$= \$100,000 + (\$100,000)\left(\frac{(1+0.06)^{19} - 1}{(0.06)(1+0.06)^{19}}\right)$$
$$= \$100,000 + (\$100,000)(11.1581)$$
$$= \$1,215,810 \quad (\$1,220,000)$$

The answer is C.

Problem 2

Joe wants to be a millionaire. To achieve this goal, at the end of the year he invests $5000 each year into an account that pays 10% interest, compounded yearly. The amount of time it will take Joe to reach his goal of becoming a millionaire is most nearly

    (A) 20 yr
    (B) 30 yr
    (C) 40 yr
    (D) 180 yr

Solution

Use the uniform series compound amount factor to relate the future worth of the account to the yearly deposits.

$$A = \text{uniform series of end of}$$
$$\text{compounding period cash flows}$$
$$= \$5000$$

$$i = \text{annual interest rate}$$
$$= 10\%$$

$$n = \text{number of compounding periods}$$
$$= \text{number of years}$$

$$F = \text{future equivalent value of a cash flow}$$
$$\text{or series of cash flows}$$
$$= \$1,000,000$$

$$\$1,000,000 = (\$5000)(F/A, 10\%, n)$$
$$= (\$5000)\left(\frac{(1+0.1)^n - 1}{0.1}\right)$$
$$21 = 1.1^n$$
$$\log 21 = n \log 1.1$$
$$n = \frac{\log 21}{\log 1.1} = 31.94 \text{ yr} \quad (30 \text{ yr})$$

The answer is B.

Problem 3

A credit card company offers students a credit line of $2000 and charges an annual percentage rate of 12%, compounded daily. The effective annual interest rate is most nearly

(A) 3.3%
(B) 12.0%
(C) 12.8%
(D) 13.2%

Solution

The following equation relates a nominal interest rate to the effective annual interest rate.

$$i = \left(1 + \frac{r}{m}\right)^m - 1$$

r = nominal interest rate = 12%

m = number of compounding periods per year

   = 365 compounding periods per year

$$i = \left(1 + \frac{0.12}{365}\right)^{365} - 1 = 0.1275 \quad (12.8\%)$$

The answer is C.

Problem 4

A production company incurs fixed costs of $1,250,000 per year. The variable cost of production is $10 per unit. All units produced are sold for $30. A maximum of 130,000 units can be manufactured each year. The break-even production quantity for the company is

(A) 47,500 units/yr
(B) 50,000 units/yr
(C) 62,500 units/yr
(D) 125,000 units/yr

Solution

The break-even production quantity yields a zero total profit.

$$\text{profit} = 0 = (p - C_v)D - C_F$$

$$D = \frac{C_F}{p - C_v}$$

p = selling price = $30/unit

C_v = variable production cost = $10/unit

C_F = fixed production costs = $1,250,000/yr

D = production quantity (units/yr)

$$= \frac{1,250,000 \ \dfrac{\$}{\text{yr}}}{30 \ \dfrac{\$}{\text{unit}} - 10 \ \dfrac{\$}{\text{unit}}}$$

$$= 62,500 \ \text{units/yr}$$

The answer is C.

Problem 5

The cost data related to the production of a new product is the following: direct labor is 0.25 hr/unit at $17/hr, direct material costs are $300 per 50 units, and overhead is 150% of total direct costs.

If the company desires to make a profit equal to 12% of the total manufacturing cost, the selling price should most nearly be set to

(A) $17/unit
(B) $25/unit
(C) $27/unit
(D) $29/unit

Solution

$$\frac{\text{selling price}}{\text{unit}} = \frac{\text{total manufacturing cost}}{\text{unit}} + \frac{\text{profit}}{\text{unit}}$$

$$\frac{\text{total manufacturing cost}}{\text{unit}} = \frac{\text{direct labor cost}}{\text{unit}} + \frac{\text{direct material cost}}{\text{unit}} + \frac{\text{overhead cost}}{\text{unit}}$$

$$\frac{\text{direct labor cost}}{\text{unit}} = \left(0.25 \ \frac{\text{hr}}{\text{unit}}\right)\left(17 \ \frac{\$}{\text{hr}}\right) = \$4.25/\text{unit}$$

$$\frac{\text{direct material cost}}{\text{unit}} = \frac{\$300}{50 \ \text{units}} = \$6/\text{unit}$$

$$\frac{\text{overhead cost}}{\text{unit}} = (150\%) \left(\frac{\text{direct labor cost}}{\text{unit}} + \frac{\text{direct material cost}}{\text{unit}} \right)$$

$$= (1.5) \left(4.25 \, \frac{\$}{\text{unit}} + 6 \, \frac{\$}{\text{unit}} \right)$$

$$= \$15.375/\text{unit}$$

$$\frac{\text{total manufacturing cost}}{\text{unit}} = 4.25 \, \frac{\$}{\text{unit}} + 6 \, \frac{\$}{\text{unit}} + 15.375 \, \frac{\$}{\text{unit}}$$

$$= \$25.625/\text{unit}$$

$$\frac{\text{profit}}{\text{unit}} = (12\%) \left(\frac{\text{total manufacturing cost}}{\text{unit}} \right)$$

$$= (0.12) \left(25.625 \, \frac{\$}{\text{unit}} \right)$$

$$= \$3.075/\text{unit}$$

$$\frac{\text{selling price}}{\text{unit}} = 25.625 \, \frac{\$}{\text{unit}} + 3.075 \, \frac{\$}{\text{unit}}$$

$$= \$28.70/\text{unit} \quad (\$29/\text{unit})$$

The answer is D.

Problem 6

A company estimates that the annual cost for a project will be \$25,000 per year for the next 10 years. The present worth equivalent of the project cost, assuming the annual interest rate of 8% is compounded quarterly, is most nearly

(A) \$166,000
(B) \$168,000
(C) \$225,000
(D) \$250,000

Solution

The effective annual interest rate with non-annual compounding is

$$i = \left(1 + \frac{r}{m} \right)^m - 1$$

$r = \text{nominal annual interest rate} = 0.08$

$m = \text{number of compounding periods per year} = 4$

$$i = \left(1 + \frac{0.08}{4} \right)^4 - 1$$

$$= 0.0824$$

The present worth equivalent of a uniform series is

$$(P/A, i, n) = A \left(\frac{(1+i)^n - 1}{i(1+i)^n} \right)$$

$$A = (\$25,000) \left(\frac{(1+0.0824)^{10} - 1}{(0.0824)(1+0.0824)^{10}} \right)$$

$$= \$165,951 \quad (\$166,000)$$

The answer is A.

Problem 7

A company purchases a machine for \$100,000 that has an expected useful life of 10 years and a salvage value of \$10,000. If the operation and maintenance costs are estimated to be \$5000 per year over the machine's life, and assuming an interest rate of 10% per year, the annual equivalent cost of owning the machine is most nearly

(A) \$16,000
(B) \$17,000
(C) \$21,000
(D) \$40,000

Solution

Both the initial cost and salvage value must be annualized over the machine's 10 year useful life.

$$A_P = P(A/P, i, n) = P \left(\frac{i(1+i)^n}{(1+i)^n - 1} \right)$$

$$= (\$100,000) \left(\frac{(0.10)(1+0.10)^{10}}{(1+0.10)^{10} - 1} \right)$$

$$= \$16,274.54$$

$$A_F = F(A/F, i, n) = F \left(\frac{i}{(1+i)^n - 1} \right)$$

$$= (\$10,000) \left(\frac{0.10}{(1+0.10)^{10} - 1} \right)$$

$$= \$627.45$$

$$\frac{\text{total annual}}{\text{equivalent}} = A_P - A_F + \frac{\text{operation and}}{\text{maintenance}}$$

$$= \$16,274.54 - \$627.45 + \$5000$$

$$= \$20,647.09 \quad (\$21,000)$$

The answer is C.

Problem 8

A company is considering purchasing a machine that costs $40,000, has an annual operation and maintenance cost of $4000, has a useful life of 10 years, and has no expected salvage value. If the revenues generated from this machine are expected to be $12,000 per year, the rate of return for this project is most nearly

    (A) 10%
    (B) 15%
    (C) 20%
    (D) 27%

Solution

The cash flows are

$$\text{year } 0 = -\$40{,}000$$
$$\text{years } 1\text{--}10 = \text{revenue} - \text{cost} = \$12{,}000 - \$4000$$
$$= \$8000$$

The present value is zero when the rate of return interest rate is used.

$$P = -\$40{,}000 + \$8000(P/A, i, 10) = 0$$
$$(P/A, i, 10) = \frac{\$40{,}000}{\$8000}$$
$$= 5.0$$

Interest tables can be used to determine the interest rate that sets $(P/A, i, 10)$ equal to 5.0. Scanning the $(P/A, i, 10)$ column of the tables, a 15% table sets $(P/A, i, 10)$ equal to 5.0. Therefore, the rate of return is 15%, and $(P/A, 15\%, 10)$.

If a 15% interest table is not available, bound the factor and interpolate. For example, if only 12% and 18% tables are available, their respective values of $(P/A, i, 10)$ are 5.65 and 4.49. Set up the interpolation formula, where the interpolation formula is defined as

interpolated value = initial value
    + (final value − initial value)
    × (ratio of interpolation increment
        divided by interpolation range)

Approximate the interest rate as

$$i = 12.0\% + (18.0\% - 12.0\%)\left(\frac{5.65 - 5.0}{5.65 - 4.49}\right)$$
$$= 12.0\% + (6.0\%)(0.56)$$
$$= 15.36\% \quad (15\%)$$

The answer is B.

Problem 9

The Missouri Department of Transportation built a new bridge over the Mississippi River for a cost of $150 million. It is estimated that the cost to maintain the bridge will be $1.5 million per year indefinitely. At a 5% interest rate, how much does the department need to invest in order to cover the bridge's perpetual maintenance cost?

    (A) $1,500,000
    (B) $15,000,000
    (C) $20,000,000
    (D) $30,000,000

Solution

This is a capitalized cost problem.

$$P = \frac{A}{i}$$
$$= \frac{\$1{,}500{,}000}{0.05}$$
$$= \$30{,}000{,}000$$

The answer is D.

PROBABILITY AND STATISTICS

Problem 10

A comparative study is made for two heat treatment processes. The results of hardness testing (in Rockwell B hardness) of ten sample pairs are as follows.

specimen	1	2	3	4	5	6	7	8	9	10
process 1	63	62	69	65	64	67	63	63	64	68
process 2	62	64	69	64	65	66	63	65	63	68

The differences in the individual pairs, d_i, are normally distributed. Calculate the test statistics and conclude whether these two processes produce different results based on a 5% significance level.

(A) $t_0 = -0.26$, so there is no evidence to indicate that these two processes produce different hardness readings.

(B) $t_0 = -0.10$, so there is no evidence to indicate that these two processes produce different hardness readings.

(C) $t_0 = 1.62$, so there is significant evidence to indicate that these two processes produce different hardness readings.

(D) $t_0 = 2.26$, so there is significant evidence to indicate that these two processes produce different hardness readings.

Solution

The pair differences ($d_i = x_{1,i} - x_{2,i}$) are $1, -2, 0, 1, -1, 1, 0, -2, 1, 0$.

$$\sum d_i = -1$$

$$\sum d_i^2 = 13$$

$$n = 10$$

The mean difference is

$$\bar{d} = \frac{1}{n}\sum d_i = \left(\frac{1}{10}\right)(-1) = -0.10$$

The estimate of standard deviation is

$$s = \sqrt{\frac{\sum d_i^2 - \frac{1}{n}\left(\sum d_i\right)^2}{n-1}} = \sqrt{\frac{13 - \left(\frac{1}{10}\right)(-1)^2}{10-1}}$$

$$= 1.20$$

$$t_0 = \frac{\bar{d}}{\frac{s}{\sqrt{n}}} = \frac{-0.10}{\frac{1.20}{\sqrt{10}}} = -0.26$$

Since the question asks if the two processes are different, a two-way (two-tail) test is required, and the percentage of exceedance allocated to each tail is $5\%/2 = 2.5\%$.

From the t-distribution table with $(n-1)$ degrees of freedom, $t_{9,1-0.025}^* = -2.262$. Since $|-0.26| < |-2.262|$, there is no evidence to indicate that these two processes produce different hardness readings.

The answer is A.

Problem 11

A one-way factorial design is used to determine whether the sample effect is significant at a 5% level. The analysis of the variance table from experimental results is shown as follows.

source of variation	degrees of freedom	sum of squares
between samples	1	4623
within samples	20	8120

The conclusions are

(A) $F = 4061/4623 = 0.88$, so the sample effect is not significant at a 5% level.

(B) $F = 4623/8120 = 0.57$, so the sample effect is not significant at a 5% level.

(C) $F = 8120/4623 = 1.76$, so the sample effect is significant at a 5% level.

(D) $F = 4623/406 = 11.39$, so the sample effect is significant at a 5% level.

Solution

From the F-distribution table, $F_{1,20,0.05}^* = 4.35$.

$$F = \frac{\text{mean of squares between samples}}{\text{mean of squares within samples}} = \frac{\dfrac{4623}{1}}{\dfrac{8120}{20}}$$

$$= 11.39$$

The sample effect is significant at a 5% level; that is, the differences between the samples are more significant than the fluctuation within the samples.

The answer is D.

Problem 12

A two-way factorial design is used to determine whether the treatment is significant at a 5% level. The analysis of the variance table from experimental results is shown as follows.

source of variation	degrees of freedom	sum of squares
replications	4	18.56
treatments	4	36.92
residuals	16	116.08

The conclusions are

(A) $F = 116.08/36.92 = 3.14$, so the treatment is significant at a 5% level.

(B) $F = 116.08/18.56 = 6.29$, so the treatment is significant at a 5% level.

(C) $F = 7.255/9.23 = 0.786$, so the treatment is not significant at a 5% level.

(D) $F = 9.23/7.255 = 1.27$, so the treatment is not significant at a 5% level.

Solution

From the F-distribution table, $F^*_{4,16,0.05} = 3.01$.

$$F = \frac{\text{mean of squares between samples}}{\text{mean of squares within samples}} = \frac{\dfrac{36.92}{4}}{\dfrac{116.08}{16}}$$

$$= 1.27$$

The sample effect from treatment is not significant at a 5% level; that is, the fluctuation within the samples is more significant than the differences between samples.

The answer is D.

Problem 13

Defects in the finished surface of furniture are approximately Poisson distributed with a mean of 0.015 defects/m^2. A finished bookshelf board containing 10 m^2 of finished surface is considered defective if it contains more than one defect. If shelves are packaged 50 per box, the probability that a box contains fewer than 2 defective items is most nearly

(A) 0.50

(B) 0.61

(C) 0.82

(D) 0.91

Solution

Let $x =$ the number of defects per board (10 m^2). x is Poisson distributed.

$$P(x) = \frac{\lambda^x e^{-\lambda}}{x!} \quad x = 0, 1, 2, \ldots$$

$$\lambda = \frac{\text{average number of defects}}{\text{board}}$$

$$= \left(\frac{10 \text{ m}^2}{1 \text{ board}}\right)\left(0.015 \, \frac{\text{defect}}{\text{m}^2}\right)$$

$$= 0.15 \text{ defects/board}$$

$$P\left(\frac{\text{board}}{\text{defective}}\right) = P(x > 1) = 1 - P(x \leq 1)$$

$$= 1 - P(x = 0) - P(x = 1)$$

$$= 1 - \frac{(0.15)^0 e^{-0.15}}{0!} - \frac{(0.15)^1 e^{-0.15}}{1!}$$

$$= 1 - 0.8607 - 0.1291$$

$$= 0.0102$$

$$P\left(\frac{\text{not}}{\text{defective}}\right) = 1 - 0.0102 = 0.9898$$

Let $y =$ number of defects in a box. y follows a binomial distribution.

$$P(y) = \binom{50}{y}(0.0102)^y(0.9898)^{50-y}$$

$$y = 0, 1, 2, \ldots, 50$$

$$P(y < 2) = P(y \leq 1) = P(y = 0) + P(y = 1)$$

$$= \binom{50}{0}(0.0102)^0(0.9898)^{50}$$

$$+ \binom{50}{1}(0.0102)^1(0.9898)^{49}$$

$$= 0.5989 + 0.006172$$

$$= 0.6051 \quad (0.61)$$

The answer is B.

Problem 14

A product consists of two parts that are placed end to end. Assume that the dimensions of the length of the parts are approximately normally distributed with the mean and standard deviation shown as follows.

	mean length (in)	standard deviation (in)
part A	2.65	0.12
part B	1.45	0.38

The probability that the combined length is greater than 4.35 in is approximately

(A) 0.20

(B) 0.26

(C) 0.55

(D) 0.90

Solution

Let $\mu_A =$ mean length of part A, and let $\mu_B =$ mean length of part B.

$$\sigma_A^2 = \text{variance of length of part A}$$

$$\sigma_B^2 = \text{variance of length of part B}$$

$$x_A = \text{length of part A}$$

$$x_B = \text{length of part B}$$

$$t = \text{total length} = x_A + x_B$$

$f(t)$ is normally distributed.

$$\mu_t = \mu_A + \mu_B = 2.65 + 1.45 = 4.1$$

The variance of t is

$$\sigma_t^2 = \sigma_A^2 + \sigma_B^2 = (0.12)^2 + (0.38)^2 = 0.1588$$

For $P(t > 4.35)$, the standard normal variant is

$$z = \frac{X - \mu_t}{\sqrt{\sigma_t^2}} = \frac{4.35 - 4.1}{\sqrt{0.1588}} = 0.63$$

$$P(z > 0.63) = 0.26$$

The answer is B.

Problem 15

Two different machines (A and B) are used to fill 20 oz containers of soda. The following data was collected. (All values are in oz.)

	bottles sampled	sample mean	sample variance
machine A	16	20.25	0.18
machine B	10	20.15	0.34

With 90% confidence, the endpoints of the ratio of the population variance A to population variance B are

(A) 0.73, 5.7
(B) 0.18, 1.4
(C) 0.13, 3.8
(D) 0.18, 4.8

Solution

$$\sigma_A^2 = \text{variance for machine A}$$
$$\sigma_B^2 = \text{variance for machine B}$$
$$s_A^2 = \text{sample variance for machine A}$$
$$s_B^2 = \text{sample variance for machine B}$$

Use the F-distribution to estimate σ_B^2.

$$\frac{1}{F_{0.05,9,15}} \leq \frac{\sigma_A^2 s_B^2}{\sigma_B^2 s_A^2} \leq F_{0.05,9,15}$$

$$\frac{s_A^2}{(F_{0.05,9,15})s_B^2} \leq \frac{\sigma_A^2}{\sigma_B^2} \leq \frac{s_A^2(F_{0.05,9,15})}{s_B^2}$$

$$\frac{0.18}{(0.34)(3.01)} \leq \frac{\sigma_A^2}{\sigma_B^2} \leq \left(\frac{0.18}{0.34}\right)(2.59)$$

$$0.176 \leq \frac{\sigma_A^2}{\sigma_B^2} \leq 1.371$$

The answer is B.

Problem 16

The following information was gathered from two sets of experiments comparing entity time-in-system (T) for systems A and B. d_i is the individual replication difference.

replication	T_A	T_B	$d_i = T_B - T_A$
1	9.19	8.87	−0.32
2	19.18	31.52	12.34
3	12.30	14.14	1.84
4	13.04	14.11	1.07
5	17.79	16.72	−1.07
6	11.49	27.78	16.29
7	21.61	34.05	12.44
8	10.50	22.96	12.46
9	13.27	10.98	−2.29
10	8.50	11.66	3.16

The average difference between the two systems is $\overline{d} = 5.59$. The standard deviation of the difference is $s(d) = 6.96$. The endpoints of the paired-t 95% confidence interval comparing time-in-system for system B with that for system A are most nearly

(A) $[-10.13, 21.31]$
(B) $[-5.28, 5.28]$
(C) $[0.80, 10.38]$
(D) $[1.56, 9.62]$

Solution

$\alpha = 1 - 0.95 = 0.05$. The confidence interval endpoints are

$$[\overline{d} - h, \overline{d} + h]$$

$$h = [(t_{n-1,1-\alpha/2})][s(\overline{d})]$$

$$t_{9,0.975} = 2.26 \quad \text{[from the student-}t\text{ table]}$$

$$s(\overline{d}) = \frac{s(d)}{\sqrt{n}}$$

$$= \frac{6.96}{\sqrt{10}}$$

$$= 2.12$$

$$h = (2.26)(2.12)$$

$$= 4.79$$

The confidence interval endpoints are

$$[5.59 - 4.79, 5.59 + 4.79] = [0.80, 10.38]$$

The answer is C.

Problem 17

A data set has been gathered with 725 values measuring the number of arrivals per minute at a workstation. Calculations show that the arrivals follow a Poisson distribution with a mean of 17 min.

What distribution and parameters should be used in a simulation model for the interarrival times (i.e., times between arrivals) at the workstation?

(A) normal, with a mean of 17 min and variance of $17/\sqrt{725}$
(B) Weibull, with an alpha of 17 min and a beta of 725
(C) Poisson, with a mean of 17 min
(D) exponential, with a mean of $1/17$ min

Solution

If the number of events that occur in a fixed time interval has a Poisson distribution with mean λ, then the time between events is exponentially distributed with a mean of $1/\lambda$.

The answer is D.

Problem 18

A car part must be processed by three different operations before it can be installed. Each operation uses multiple machines to process the part. Operation 1 has seven machines, of which the part must use three. Operation 2 has five machines, of which the part must use two. Operation 3 has four machines, of which the part must use one. In each operation, the order that the machines process the part does not matter. How many different operation and machine combinations are there to completely process a part?

(A) 50
(B) 200
(C) 1400
(D) 17,000

Solution

This is a combination problem, not a permutation problem, because it does not matter what order the machines process the car part. To find the total number of combinations for each operation and machine, use the following equation, where n is the number of objects taken r at a time.

$$C(n, r) = \frac{n!}{r!(n-r)!}$$

For operation 1,

$$C(7, 3) = \frac{7!}{3!(7-3)!}$$
$$= 35$$

For operation 2,

$$C(5, 2) = \frac{5!}{2!(5-2)!}$$
$$= 10$$

For operation 3,

$$C(4, 1) = \frac{4!}{1!(4-1)!}$$
$$= 4$$

Serially multiply together the operations' numbers of combinations to find the total number of combinations.

$$\text{total number of combinations} = \prod C_i = (35)(10)(4)$$
$$= 1400$$

The answer is C.

MODELING AND COMPUTATION

Problem 19

An industrial engineer has been asked to write computer code that will determine the mean and standard deviation for the process time of a given part. The data input file to be read contains information on all parts produced in the plant. Data that is corrupted is not retained for calculation. Which of the following flow charts represents the correct logic for the computer code?

(A)

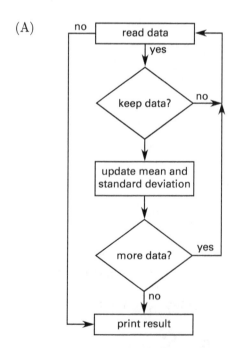

(B)

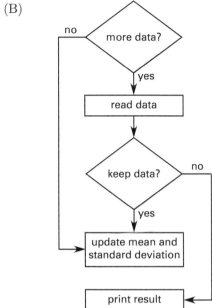

(D)

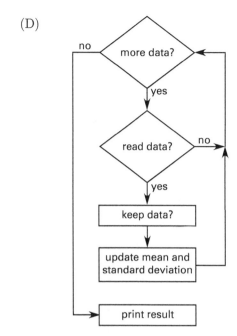

(C)

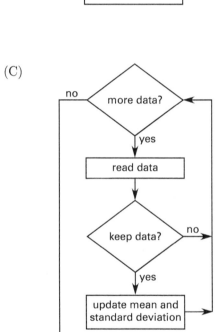

Solution

First, the program must check to see if there is more data to be read. If there is, then the data is read, otherwise the current result is printed. If data is read, then the program must determine if the data is the correct type. If it is the correct type, then the mean and standard deviation are updated, and the program looks for more data. Otherwise, the program looks for more data and skips updating.

The answer is C.

Problem 20

In an entity-relationship diagram for a relational database, the relationship between a product record and the part family record it belongs to will be most likely represented as

(A) many to one (n:1)
(B) one to many (1:n)
(C) one to one (1:1)
(D) many to many (n:m)

Solution

In group technology, a product can belong to only one part family (:1), but the part family can have many products as members of the family (n:). Thus, the relationship between the product record and the part family record it belongs to is many to one (n:1).

The answer is A.

Problem 21

For the following problem, what is the optimal solution for both the original problem and its dual? Let y represent the dual values.

$$\begin{aligned} \text{minimize} \qquad & 3x_1 + 12x_2 \\ \text{subject to} \qquad 2x_1 + 4x_2 - 2x_3 & \geq 6 \\ -6x_1 - 3x_2 + 12x_3 & \geq 6 \\ x & \geq 0 \end{aligned}$$

(A) $x_1 = 7$, $x_2 = 0$, $x_3 = 4$, and $y_1 = 3$, $y_2 = \frac{1}{2}$
(B) $x_1 = 0$, $x_2 = 2$, $x_3 = 0$, and $y_1 = \frac{1}{2}$, $y_2 = 3$
(C) $x_1 = 7$, $x_2 = 0$, $x_3 = 4$, and $y_1 = 2$, $y_2 = \frac{1}{2}$
(D) $x_1 = 1$, $x_2 = 2$, $x_3 = 2$, and $y_1 = 4$, $y_2 = \frac{1}{2}$

Solution

To be a solution, the points must satisfy the constraints (i.e., they must be feasible) of their respective problem, and the objective functions of the original and dual problems must be equal. The dual objective function for this problem is given by the right-hand side of the original problem and is $6y_1 + 6y_2$. The constraints for the dual are given by the columns and the objective function coefficients of the original problem. The dual has the form

$$\begin{aligned} \text{maximize} \qquad & 6y_1 + 6y_2 \\ \text{subject to} \qquad 2y_1 - 6y_2 & \leq 3 \\ 4y_1 - 3y_2 & \leq 12 \\ -2y_1 + 12y_2 & \leq 0 \\ y_1, y_2 & \geq 0 \end{aligned}$$

The solution given in choice (A) satisfies the constraints, and the two objective functions are equal. For the original problem constraints,

$$(2)(7) + (4)(0) - (2)(4) = 6$$
$$(-6)(7) - (3)(0) + (12)(4) = 6$$

For the dual problem constraints,

$$(2)(3) - (6)\left(\tfrac{1}{2}\right) = 3$$
$$(4)(3) - (3)\left(\tfrac{1}{2}\right) = 10.5 \leq 12$$
$$(-2)(3) + (12)\left(\tfrac{1}{2}\right) = 0$$

The objective functions are

$$(3)(7) + (12)(0) = 21 = (6)(3) + (6)\left(\tfrac{1}{2}\right)$$

Note that the solution given by choice (B) is not feasible for the original problem. The solution given by choice (C) does not yield objective function values that are equal. The solution given by choice (D) is not feasible for the dual problem.

The answer is A.

Problem 22

What is the optimal solution for the following problem?

$$\begin{aligned} \text{maximize} \qquad & 5x_1 + x_2 \\ \text{subject to} \qquad x_1 + 2x_2 & \leq 10 \\ -3x_1 + 4x_2 & \leq 10 \\ 3x_1 + 2x_2 & \leq 18 \\ x_1, x_2 & \geq 0 \end{aligned}$$

(A) $x_1 = 4$ and $x_2 = 3$
(B) $x_1 = 10$ and $x_2 = 0$
(C) $x_1 = 6$ and $x_2 = 0$
(D) $x_1 = 5$ and $x_2 = 3$

Solution

This is a simple two-dimensional problem and can be solved graphically. The optimal basic feasible solution corresponds to a vertex of the feasible region. The optimal vertex is the last feasible point that the objective function contours will touch as the value of the objective function is increased.

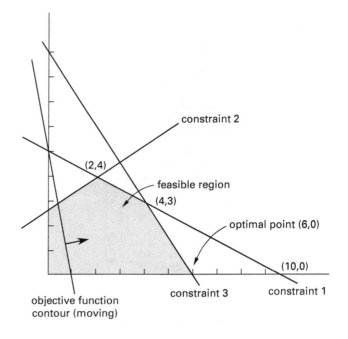

The answer is C.

Problem 23

Consider the following linear programming formulation.

$$\begin{aligned} \text{maximize} \qquad & 3x_1 + x_2 + 2x_3 \\ \text{subject to} \qquad x_1 - x_2 + 2x_3 & \leq 20 \\ 2x_1 + x_2 - x_3 & \leq 60 \\ x & \geq 0 \end{aligned}$$

Let x_4 and x_5 be the corresponding slack variables for constraints 1 and 2, respectively. The optimal tableau is

objective function coefficient		3	1	2	0	0	
c_B	x_B	x_1	x_2	x_3	x_4	x_5	b
2	x_3	3	0	1	1	1	30
1	x_2	5	1	0	1	2	40
reduced cost values	r_j	-8	–	–	-3	-4	

The most one would be willing to pay for additional units of constraint 2 is most nearly

(A) 1
(B) 2
(C) 3
(D) 4

Solution

The most one would be willing to pay for additional units of a constraint is based upon the shadow price of that constraint. The shadow price comes from the dual value associated with the constraint. In the optimal tableau, the dual values can be obtained directly from the reduced costs associated with the slack variables. In this problem, x_5 is the slack variable associated with the second constraint. Therefore, the "value" of the second constraint is 4. This means that for each increase of 1 unit of the second constraint, the objective function will increase by 4. Therefore, as long as each additional unit of the second constraint costs less than 4, it would be advantageous to purchase those units.

The answer is D.

Problem 24

A national sports team uses an automated answering service for taking ticket orders. The service is always in one of two modes: answering the phone, or off the phone. Calls come in at the rate of 10 per hour. The service can handle 15 calls per hour. Due to a high demand for tickets, the service runs 24 hours a day. The probability that the service will be answering a call three weeks from now is most nearly

(A) 0.30
(B) 0.40
(C) 0.50
(D) 0.60

Solution

This is an example of a continuous Markov chain. Since the rates are per hour and the service runs 24 hours a day, three weeks from now represents a long-term horizon. Therefore, to find the probability that the service will be on the phone, one only needs to find the steady-state probability for being on the phone. This is done through solving the system $\boldsymbol{\pi\Lambda} = 0$ and $\Sigma\pi_i = 1$, where $\boldsymbol{\Lambda}$ is the transition rate matrix and $\boldsymbol{\pi}$ is the steady-state probability vector. The elements λ_{ij}, $i \neq j$, of the matrix $\boldsymbol{\Lambda}$ are the transition rates from state i to state j. The diagonal elements are given by $\lambda_{ii} = -(\Sigma_{j \neq i}\lambda_{ij})$. Here π_1 represents the probability of being in state 1 and π_2 represents the probability of being in state 2. For this problem, if state 1 is "off the phone" and state two is "on the phone," the transition rate matrix is

$$\boldsymbol{\Lambda} = \begin{pmatrix} -10 & 10 \\ 15 & -15 \end{pmatrix}$$

The steady-state equations would be

$$(\pi_1 \ \pi_2) \begin{pmatrix} -10 & 10 \\ 15 & -15 \end{pmatrix} = (0 \ 0)$$

This yields the following set of equations.

$$-10\pi_1 + 15\pi_2 = 0$$
$$10\pi_1 - 15\pi_2 = 0$$

This system reduces to $-10\pi_1 + 15\pi_2 = 0$, which has the solution

$$\pi_1 = \tfrac{3}{2}\pi_2$$

By the normalizing equation ($\Sigma\pi_i = 1$),

$$\tfrac{3}{2}\pi_2 + \pi_2 = 1$$

This implies $\pi_2 = \tfrac{2}{5} = 0.4$.

$$\pi_1 = \tfrac{3}{5} = 0.6$$

Therefore, the probability of being on the phone (state 2) is $1 - 0.6 = 0.4$.

The answer is B.

Problem 25

For the following transition diagram of a discrete Markov chain, identify whether the states are transient, absorbing, or recurrent.

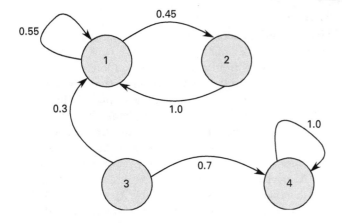

(A) All states are recurrent.
(B) State 1 is recurrent, states 2 and 3 are transient, and state 4 is absorbing.
(C) State 4 is absorbing, and states 1, 2, and 3 are transient.
(D) States 1 and 2 are recurrent, state 3 is transient, and state 4 is absorbing.

Solution

Recurrent means that a state communicates with itself. That is, once it leaves the state it will eventually return to the state. *Transient* means that eventually it will never be able to return to the state. An *absorbing* state is recurrent and has a one-step transition probability of 1.0 of going from itself to itself. From these definitions, state 4 is absorbing since its one-step transition probability is 1.0. State 3 is transient since a transition from state 3 either takes it to the absorbing state 4, or to a closed system of states 1 and 2 from which it is impossible to return to state 3. States 1 and 2 are recurrent since there is a possibility of returning to each of these states.

The answer is D.

Problem 26

The emergency room at a small local hospital has only one doctor on duty. The facility has three examination rooms and a waiting room that can hold seven people. Patients arrive according to a Poisson distribution with an arrival rate of ten people per hour. The doctor can examine and treat a patient according to an exponential distribution with a mean of 4 minutes. The average time a patient waits before seeing the doctor is most nearly

(A) 4.0 min
(B) 4.5 min
(C) 7.3 min
(D) 20 min

Solution

This is an M/M/1/10 model. The service and arrival times are exponentially distributed. There is one server, and the capacity of the system is 10 (9 people waiting and 1 being served). The average waiting time before a patient sees a doctor is given by W_q. By Little's law, $W_q = L_q/\lambda$.

For this system, the arrival rate, λ, is 10 per hour, and the service rate, μ, is 15 per hour.

$$\rho = \frac{\lambda}{\mu} = \frac{10 \ \dfrac{\text{people}}{\text{hr}}}{15 \ \dfrac{\text{people}}{\text{hr}}} = \frac{2}{3}$$

To find W_q, find $L_q = L - (1 - P_0)$.

$$P_0 = \frac{1 - \rho}{1 - \rho^{10+1}} = \frac{\frac{1}{3}}{1 - \left(\frac{2}{3}\right)^{11}} = 0.3372$$

$$L = \frac{\rho}{1 - \rho} - \frac{(10 + 1)\rho^{10+1}}{1 - \rho^{10+1}} = 2 - \frac{(11)\left(\frac{2}{3}\right)^{11}}{1 - \left(\frac{2}{3}\right)^{11}}$$

$$= 1.87135$$

$$L_q = 1.87135 - (1 - 0.3372) = 1.20855$$

$$W_q = \left(\frac{1.20855}{10 \ \dfrac{\text{people}}{\text{hr}}}\right) (60 \ \text{min})$$

$$= 7.25 \ \text{min} \quad (7.3 \ \text{min})$$

The answer is C.

Problem 27

Simulation experiments modeling a nonterminating system are often biased by the empty-and-idle start-up conditions of the experiment which do not reflect the system's steady-state operating conditions.

Which of the following is NOT a viable method for dealing with initial condition bias when experimenting with a model of a nonterminating system?

(A) Discard statistics gathered during the start-up phase.
(B) Perform a statistically significant number of replications of the experiment to overcome the start-up bias.
(C) Run the experiment for a sufficiently long period such that the steady-state statistics effectively overwhelm those from the start-up phase.
(D) Initialize the experiment states and conditions such that the start-up conditions are representative of those of steady state.

Solution
Replicating an experiment simply produces multiple data sets, each of which contain biases due to the initial conditions.

The answer is B.

Problem 28
When running a simulation experiment, data are gathered on output Y. For each replication of the experiment, a single statistic, $\overline{Y}$ (the average of the Y values), is saved. When analyzing the data across all replications, statistical tests are applied to an aggregated statistic, $\overline{\overline{Y}}$ (the average of the $\overline{Y}$ values).

When it is desired to perform statistical tests, the reason for analyzing experimental statistics in this way is to take advantage of what statistical phenomenon?

(A) Chebyshev's theorem
(B) central limit theorem
(C) Markov's theorem
(D) Kolmogorov's theorem

Solution
The central limit theorem states that when a series of independent means (or sums) are combined into a single mean (or sum), the resulting mean (or sum) has a distribution that approaches normal as the number of samples being combined increases. Therefore, the resultant statistic can be assumed to come from an independent, identical, normally distributed data set. These assumptions are required to apply most standard statistical tests such as confidence intervals, tests of equal means, and so on.

The answer is B.

Problem 29
Simulation experiments are often repeated in what are called runs or, more commonly, replications.

Replications are used in simulation experimentation because

(A) random numbers generated by computer simulation packages are not truly random, and so the sequences of numbers they produce repeat after only a small quantity of numbers are drawn
(B) many standard statistical tests used to analyze simulation outputs assume the data to be independent and identically distributed, and statistics gathered within a single replication rarely meet those requirements

(C) bias due to the initial start-up conditions must be recognized, and replication is an effective method to eliminate such bias
(D) most simulation packages have difficulty storing the vast amounts of data that would need to be collected during a single long run if no replications were used

Replications are used in simulation experiments because many standard statistical tests used to analyze simulation outputs assume the data to be independent and identically distributed, and statistics gathered within a single replication rarely meet those requirements. The central limit theorem is applicable as it states that when a series of independent means (or sums) are combined into a single mean (or sum), the resulting mean (or sum) has a distribution that approaches normalcy as the number of samples being combined increases. Therefore, the resultant statistic can be assumed to come from an independent, identically, normally distributed data set.

The answer is B.

Problem 30
Which of the following systems would best be modeled as a nonterminating system?

(A) a bank that opens in the morning and closes after 8 hours
(B) an automotive assembly line that operates for 8 hours a day, 5 days a week
(C) a military battle between two opposing armies
(D) a retailer's inventory planning system for a holiday season

Solution
Terminating systems are those that start from an empty and idle condition and end with similar conditions. They have an innate beginning and ending. Nonterminating systems are those that do not have a clear ending condition to their operation during the experimental time horizon. In this question, choices (A), (C), and (D) have reasonably clear starting and ending conditions and, therefore, are terminating systems. The system described in choice (B), while operating 8 hours a day, 5 days a week, has no such obvious starting and ending conditions. It is unlikely that an automotive assembly line is restarted each day or each week empty and idle. Therefore, the system described in choice (B) is best modeled as a nonterminating system.

Problem 31

What is the dual of the following problem?

$$\begin{array}{ll} \text{maximize} & 5x_1 + 3x_2 \\ \text{subject to} & -2x_1 + 2x_2 \le 8 \\ & 3x_1 - 6x_2 \le 10 \\ & x_1,\ x_2 \ge 10 \end{array}$$

(A) $\begin{array}{ll} \text{minimize} & 5y_1 + 3y_2 \\ \text{subject to} & -2y_1 - 2y_2 \ge 8 \\ & -3y_1 + 6y_2 \ge 10 \\ & y_1,\ y_2 \ge 0 \end{array}$

(B) $\begin{array}{ll} \text{minimize} & 8y_1 + 10y_2 \\ \text{subject to} & -2y_1 + 3y_2 \ge 5 \\ & 2y_1 - 6y_2 \ge 3 \\ & y_1,\ y_2 \ge 0 \end{array}$

(C) $\begin{array}{ll} \text{maximize} & 8y_1 + 10y_2 \\ \text{subject to} & -2y_1 + 2y_2 \ge 5 \\ & -3y_1 + 6y_2 \ge 3 \\ & y_1,\ y_2 \ge 0 \end{array}$

(D) $\begin{array}{ll} \text{maximize} & 8y_1 + 10y_2 \\ \text{subject to} & -2y_1 + 3y_2 \le 5 \\ & 2y_1 - 6y_2 \le 3 \\ & y_1,\ y_2 \ge 0 \end{array}$

Solution

The original problem is called the *primal problem*. In formulating the dual, there is one dual variable for each primal constraint. The right-hand-side values of the primal form the optimization function coefficients for the dual, and the optimization function coefficients for the primal form the right-hand-side values for the dual. Likewise, the columns of the constraint matrix of the primal are the rows of the constraint matrix for the dual. When the primal problem is in what is called *canonical* form, the dual is easy to formulate. The canonical form is dependent on the sense of optimization (max or min). The canonical forms of the primal problem and their corresponding dual problems are given as follows. For these problems, P stands for the primal problem, D for the dual problem, and s.t. for subject to.

(P) $\max \ cx$
  s.t. $Ax \le b$
    $x \ge 0$

(D) $\min \ b^T y$
  s.t. $A^T y \ge 0$
    $y \ge 0$

(P) $\min \ cx$
  s.t. $Ax \ge b$
    $x \ge 0$

(D) $\max \ b^T y$
  s.t. $A^T y \le 0$
    $y \ge 0$

The problem stated is of the canonical form given by the first primal and dual problem. Therefore, the dual is given by choice (B).

The answer is B.

Problem 32

Interpret the information contained in the following decision table.

HOLE	X
TOLERANCE > 0.01	
0.01 μ, TOLERANCE > 0.002	
TOLERANCE [0.002	X
drill	X
semi-finish bore	X
finish bore	X

(Note: TOLERANCE [0.002 implies that the hole can be greater than specified by at most 0.002, but not less than specification. 0.01 μ is the surface finish.)

(A) If the part has a hole and tolerance [0.002, then drill and semi-finish bore.

(B) If the part has a hole and tolerance [0.002, then drill, semi-finish bore, and finish bore.

(C) If the part has a hole and 0.002 < tolerance, 0.01 μ, then drill, semi-finish bore, and finish bore.

(D) If the part has a hole and tolerance > 0.01, then drill and finish bore.

Solution

A decision table summarizes different IF-THEN conditional information. The first column lists the IF conditions in the shaded top rows and the THEN consequences in the unshaded bottom rows. The columns to the right list the different production rules. The table in Prob. 32 shows one production rule. The pertinent IF-THEN statements are denoted by an X mark. The production rule states that IF there is a hole feature with a tolerance [0.002, THEN use drill, semi-finish bore, and finish bore operations.

The answer is B.

Problem 33

A G/M/2/4 queue always has

(A) exponentially distributed interarrival times
(B) normally distributed service times
(C) two busy servers
(D) less than five customers

Solution

According to the Kendall notation, a G/M/2/4 queue has generally distributed interarrival times (G) (which could be any distribution), exponentially distributed services times (M), two servers who are either busy or idle, and up to four customers.

The answer is D.

INDUSTRIAL MANAGEMENT

Problem 34

Motivating workers to perform efficiently and effectively has long been a problem facing industrial managers. A significant 1940s research project introduced what is called the "hierarchy-of-needs theory." As shown in the following illustration, this theory recognized that people are motivated by five distinct types of needs: physiological, safety, love, esteem, and self-actualization.

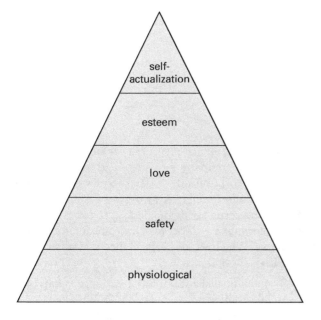

The researcher whose name is now synonymous with the hierarchy-of-needs theory is

(A) Frederick W. Taylor
(B) Frank B. Gilbreth
(C) Peter F. Drucker
(D) Abraham H. Maslow

Solution

Abraham H. Maslow first published his hierarchy-of-needs theory in the journal *Psychological Review* in 1943.

The answer is D.

Problem 35

A recent industrial engineering graduate was hired by a large aircraft manufacturing company. The engineer works in the systems engineering department where the department manager ensures that the engineer has the necessary systems tools and contacts with other systems experts to perform her job. Shortly after being hired, the engineer is assigned to a large project involving the design and manufacture of a new commercial cargo plane. The cargo plane project manager employs the engineer's skills in combination with the skills from other disciplines to tackle the project. The engineer has, in essence, two bosses: the department manager and the project manager.

The organizational structure the engineer is experiencing at the aircraft company could best be described as a

(A) functional organization
(B) line-staff organization
(C) matrix organization
(D) network organization

Solution

A matrix organizational structure applies dual chains of command. Functional departmentalization is used to gain economies from specialization. Overlaying the functional departments is a set of managers responsible for specific projects or products.

The answer is C.

Problem 36

Business process reengineering (BPR) has been defined as the fundamental analysis and radical redesign of strategic, value-added business processes, and the systems, policies, and organizational structures that support them, which results in a dramatic improvement in critical measures of business performance. In fact, some have labeled BPR as "the modern definition of Industrial Engineering." Yet, recent studies have shown that as few as one-third of BPR efforts are successful.

Experts in the field of BPR have identified ten primary reasons why BPR efforts fail. Which of the following is NOT a primary reason for failures in BPR efforts?

 (A) BPR work can conflict with team members' "real jobs" or with other improvement programs.
 (B) BPR team members lack the engineering skills to design and implement large-scale system changes.
 (C) BPR efforts are attempted without defined methodologies.
 (D) BPR teams consist of only selected representatives rather than people from all levels of all affected departments.

Solution

Unlike continuous improvement programs, BPR involves radical and fundamental change. While it is essential to consider the human dimension of such radical and fundamental change, it is generally thought that involving representatives from all levels of all affected departments is a mistake. To use a medical analogy, continuous improvement is like vitamins and exercise while BPR is like radical surgery. Such surgery is best performed by a small, skilled team rather than an all-encompassing committee.

The answer is D.

Problem 37

Which of the following lists only contains characteristics of successful incentive systems?

 (A) ceilings on employee earnings, increased unit cost, and gain sharing with employer
 (B) increased rate of production, fair standards, reduced unit cost, and increased employee earnings
 (C) complicated pay formulas, good standards, and selective application to employees
 (D) increased employee earnings, absence of union support, and unchanged rate of production

Solution

A successful incentive system provides benefits to both management and the employee. Therefore, the system would be expected to increase production and associated productivity (which would reduce cost), be fair, be

understandable to employees, and increase their earning potential.

The answer is B.

Problem 38

A company uses a piecework incentive system. The base rate is $12.00 per hour for a 10 hr work day. For a particular assembly operation, the standard time per piece is 0.6 min. What would a worker be paid for assembling 1150 pieces in one day?

 (A) $115
 (B) $123
 (C) $138
 (D) $157

Solution

The piecework rate is

$$(\text{base rate}) \left(\frac{\text{standard time}}{\text{pieces}} \right)$$

$$= \left(12.00 \ \frac{\$}{\text{hr}} \right) \left(0.6 \ \frac{\text{min}}{\text{piece}} \right) \left(\frac{1}{60} \ \frac{\text{hr}}{\text{min}} \right)$$

$$= \$0.12/\text{piece}$$

The pay is

$$(\text{piecework rate})(\text{no. of pieces completed})$$

$$= (\$0.12)(1150)$$

$$= \$138$$

The answer is C.

Problem 39

Traditional performance measures are not always sufficient to guide a company into sustained success. Issues that motivate corporations to develop new performance measures include fewer overall organizational levels, a need for long-term strategic perspective, an over reliance on accounting data, and a desire for more simple systems. Which set of performance measures would best address these issues?

 (A) machine utilization, labor variance, scrap cost, labor efficiency
 (B) cycle time, sales, on-time delivery, customer satisfaction, inventory turns
 (C) daily production, overhead variance, return on investment, inventory level
 (D) labor variance, machine efficiency, payback, daily production

Solution

Performance measures relevant to current manufacturers need to be directly related to strategy, as well as be primarily non-financial, simple and easy to use, and be able to do more than just monitor performance by providing useful feedback to managers. Measures such as cycle time, sales, on-time delivery, customer satisfaction, and inventory turns provide managers with the ability to make meaningful decisions. The other options, such as utilization, efficiency, budget variance, and so forth, do not.

The answer is B.

MANUFACTURING AND PRODUCTION SYSTEMS

Problem 40

A single-point tool is used on a lathe to make a single pass cut on a 2 in diameter (before the cut), 24 in long part. The depth of cut is set to 0.125 in at a spindle speed of 400 rpm with a feed of 0.012 in per revolution.

The metal removal rate is most nearly

    (A) $0.04 \text{ in}^3/\text{min}$
    (B) $3.5 \text{ in}^3/\text{min}$
    (C) $15 \text{ in}^3/\text{min}$
    (D) $58 \text{ in}^3/\text{min}$

Solution

Find the metal removal rate (MRR).

$$A = \text{cross-sectional area}$$
$$V = \text{volume}$$
$$t = \text{time}$$
$$f = \text{feed}$$
$$s = \text{spindle speed}$$
$$r_o = \text{outer radius}$$
$$r_i = \text{inner radius}$$

$$\text{MRR} = \frac{V}{t} = Afs = \pi(r_o^2 - r_i^2)fs$$
$$= \pi\left((1.000 \text{ in})^2 - (0.875 \text{ in})^2\right)$$
$$\times \left(0.012 \frac{\text{in}}{\text{rev}}\right)\left(400 \frac{\text{rev}}{\text{min}}\right)$$
$$= 3.534 \text{ in}^3/\text{min} \quad (3.5 \text{ in}^3/\text{min})$$

The answer is B.

Problem 41

Using the production flow analysis method (or any other equivalent clustering procedure), determine the appropriate machine cell and part family grouping for the following machine/part matrix.

	C1	C2	C3	C4	C5
M1		1		1	
M2	1	1	1	1	1
M3		1		1	
M4	1		1		1
M5	1				1

    (A) family/cell A (C2, C4/ M2, M4, M5), family/cell B (C1, C3, C5/ M1, M3)
    (B) family/cell A (C2, C4/ M1, M3), family/cell B (C1, C3, C5/ M2, M4, M5)
    (C) family/cell A (C1, C5/ M2, M4, M5), family/cell B (C2, C3, C4/ M1, M2, M3, M4)
    (D) family/cell A (C1, C3, C5/ M2, M4, M5), family/cell B (C2, C3, C4/ M1, M2, M3)

Solution

Using Burbridge's production flow analysis method that assigns binary weights to the rows and columns iteratively until ordering is achieved (or using any other equivalent clustering method such as similarity coefficient weighting) will ultimately lead to a grouping of machines and parts as given in choice (B).

	C1	C2	C3	C4	C5	
M1		1		1		2^0
M2	1	1	1	1	1	2^1
M3		1		1		2^2
M4	1		1		1	2^3
M5	1				1	2^4
	26	7	10	7	26	

	C2	C4	C3	C1	C5	
M1	1	1				3
M2	1	1	1	1	1	31
M3	1	1				3
M4			1	1	1	28
M5				1	1	24
	2^0	2^1	2^2	2^3	2^4	

	C2	C4	C3	C1	C5	
M1	1	1				2^0
M3	1	1				2^1
M5				1	1	2^2
M4			1	1	1	2^3
M2	1	1	1	1	1	2^4
	19	19	24	28	28	

The answer is B.

Problem 42

Jobs go through a drilling station and then an automatic inspection station. Each station can only service one job at a time. Four jobs are currently awaiting processing. The time each job must spend at each station is shown as follows.

job	drilling time	inspection time
1	12	9
2	8	5
3	6	8
4	3	4

The job sequence that minimizes the makespan schedule is

(A) $\{4,3,2,1\}$
(B) $\{4,2,1,3\}$
(C) $\{4,3,1,2\}$
(D) $\{4,2,3,1\}$

Solution

Johnson's rule results in the minimum makespan schedule for a two-machine flow shop. The following steps detail the application of Johnson's rule to the previous problem.

The shortest time is the drilling operation of job 4. Since this time is at the first workstation, job 4 is placed first in the sequence.

current sequence | 4 | | | |

The shortest time of remaining jobs is the inspection time of job 2. Since this time is at the second station, job 2 is placed last in the sequence.

current sequence | 4 | | | 2 |

The shortest time of remaining jobs is now the drilling operation of job 3. Since this time is at the first station, job 3 is placed behind job 4 at the beginning of the sequence.

current sequence | 4 | 3 | | 2 |

Only job 1 remains. Job 1 is placed in the remaining slot in the sequence.

final sequence | 4 | 3 | 1 | 2 |

The answer is C.

Problem 43

The following table contains sales data for product X over the past twelve months.

month	units sold	month	units sold
1	100	7	120
2	93	8	108
3	102	9	83
4	78	10	96
5	85	11	115
6	98	12	110

Using a three-month moving average, the forecast for sales of product X during the next month is most nearly

(A) 99 units
(B) 107 units
(C) 110 units
(D) 112 units

Solution

The general equation for a moving average forecast is

$$F_t = \frac{1}{n}\sum_{i=1}^{n} D_{t-i}$$

F_t = forecast for period t

D_t = actual demand in period t

n = number of most recent observations to include in the forecast for the next period

For this problem, $t = 13$ and $n = 3$.

$$F_{13} = \frac{1}{3}\sum_{i=1}^{3} D_{13-i} = \left(\tfrac{1}{3}\right)(D_{12} + D_{11} + D_{10})$$

$$= \left(\tfrac{1}{3}\right)(110 \text{ units} + 115 \text{ units} + 96 \text{ units})$$

$$= 107 \text{ units}$$

The answer is B.

Problem 44

Five jobs are waiting to be processed at a machining center. The processing times and due dates (number of days remaining until job is due) for these jobs are as follows.

job number	processing time (days)	days remaining until job is due
1	2.00	4
2	1.50	8
3	4.00	5
4	3.75	6
5	0.75	2

What is the job sequence that minimizes the average time a job will spend waiting to be processed?

(A) $\{3, 5, 1, 4, 2\}$
(B) $\{5, 2, 1, 4, 3\}$
(C) $\{5, 1, 3, 4, 2\}$
(D) $\{3, 4, 1, 2, 5\}$

Solution

Sequence the jobs in nondecreasing order of processing time (shortest processing time rule) to minimize the mean waiting time at the machining center.

job number	processing time
5	0.75
2	1.50
1	2.00
4	3.75
3	4.00

The sequence that minimizes the mean waiting time is $\{5, 2, 1, 4, 3\}$.

The answer is B.

Problem 45

A newly hired industrial engineer has been asked to conduct a lean manufacturing assessment of a key production line. In order to better understand the line, the engineer constructs the following current value stream map.

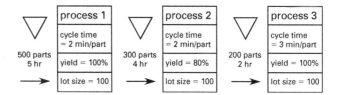

The value-added to non-value-added (VA:NVA) ratio for the current state value stream map is most nearly

(A) 0.2
(B) 0.5
(C) 1.0
(D) 1.1

Solution

Value-added activities are defined as operations that transform products that customers are willing to buy. Non-value-added activities are essentially any operations or activities that the customer is not willing to buy and are often referred to as waste. Lean manufacturing is a process management philosophy based on the Toyota Production System, which defines seven kinds of waste including overproduction, waiting, transportation waste, processing waste, inventory waste, wasted motion, and product defects.

Based on these definitions, the value-added (VA) time is determined for each process.

For process 1,

$$\left(2 \;\frac{\text{min}}{\text{part}}\right)(100 \text{ parts}) = 200 \text{ min}$$

For process 2,

$$\left(2 \;\frac{\text{min}}{\text{part}}\right)(100 \text{ parts}) = 200 \text{ min}$$

For process 3,

$$\left(3 \;\frac{\text{min}}{\text{part}}\right)(100 \text{ parts}) = 300 \text{ min}$$

total VA time $= 200$ min $+ 200$ min $+ 300$ min
$$= 700 \text{ min}$$

The non-value-added (NVA) time is determined for each process.

For inventory before process 1,

$$5 \text{ hr} = 300 \text{ min}$$

For inventory before process 2,

$$4 \text{ hr} = 240 \text{ min}$$

For defect/scrap time at process 2,

$$(0.2)\left(2 \frac{\text{min}}{\text{part}}\right)(100 \text{ parts}) = 40 \text{ min}$$

For inventory before process 3,

$$2 \text{ hr} = 120 \text{ min}$$

total NVA time $= 300$ min $+ 240$ min
$$+ 40 \text{ min} + 120 \text{ min}$$
$$= 700 \text{ min}$$

Therefore, the ratio of VA to NVA is 700 min/700 min, which is equal to 1.0.

The answer is C.

Problem 46

ACME Chemical produces a wide range of cleaning solvents, including the Super Cleaner which removes tough stains from carpet. Super Cleaner can be produced at an annual rate of 15,000 L; however, currently ACME only has an annual demand for 10,000 L. The unit cost to produce one liter of cleaner is $0.75. Before each production run of Super Cleaner, the production equipment must be cleaned and set up, which costs $40. ACME's annual inventory holding cost is 20%. What production quantity would most nearly be economical for ACME?

(A) 1800 L
(B) 3500 L
(C) 4000 L
(D) 5000 L

Solution

The formula for determining the economic production quantity is

$$Q^* = \sqrt{\frac{2AD}{h\left(1 - \dfrac{D}{R}\right)}}$$

$A =$ setup cost per lot $= \$40$
$D =$ annual demand $= 10{,}000$ L/yr
$R =$ annual replenishment rate $= 15{,}000$ L/yr
$h =$ annual holding cost per unit
$$= (0.20)\left(\$0.75 \frac{\text{L}}{\text{yr}}\right)$$
$$= \$0.15 \text{ L/yr}$$

$$Q^* = \sqrt{\frac{(2)(\$40)\left(10{,}000 \dfrac{\text{L}}{\text{yr}}\right)}{\left(\$0.15 \dfrac{\text{L}}{\text{yr}}\right)\left(1 - \dfrac{10{,}000 \dfrac{\text{L}}{\text{yr}}}{15{,}000 \dfrac{\text{L}}{\text{yr}}}\right)}}$$

$$= 4000 \text{ L}$$

The answer is C.

Problem 47

The following table contains the power demand for Columbia Electric over the past six months.

mo	demand (MW-hr)
1	3.7
2	4.1
3	3.9
4	4.2
5	4.1
6	4.3

Use exponential smoothing with a smoothing constant of 0.3 to most nearly forecast the power demand for month 7.

(A) 4.0 MW-hr
(B) 4.1 MW-hr
(C) 4.2 MW-hr
(D) 4.3 MW-hr

Solution

The equation for an exponentially weighted moving average is

$$\hat{d}_t = \text{forecasted demand for time } t$$
$$= \alpha d_{t-1} + (1-\alpha)\hat{d}_{t-1}$$
$d_t =$ actual demand for time t
$\alpha =$ smoothing constant $= 0.3$

To get started, arbitrarily set $\hat{d}_1$ to the average actual power demand for the first five months.

$$\hat{d}_1 \approx \frac{\begin{matrix} 3.7 \text{ MW-hr} + 4.1 \text{ MW-hr} + 3.9 \text{ MW-hr} \\ + 4.2 \text{ MW-hr} + 4.1 \text{ MW-hr} \end{matrix}}{5}$$
$$= 4.0 \text{ MW-hr/mo}$$

$$\hat{d}_2 = (0.3)(3.7 \text{ MW-hr}) + (1 - 0.3)(4.0 \text{ MW-hr})$$
$$= 3.91 \text{ MW-hr}$$

The remainder of the table is completed similarly.

t	forecasted $\hat{d}_t$	actual d_t	forecasted $\hat{d}_{t+1}$
1	4.0	3.7	3.91
2	3.91	4.1	3.967
3	3.967	3.9	3.9469
4	3.9469	4.2	4.023
5	4.023	4.1	4.046
6	4.046	4.3	4.122
7	4.122		

$$\hat{d}_7 = 4.122 \text{ MW-hr} \quad (4.1 \text{ MW-hr})$$

The answer is B.

FACILITIES AND LOGISTICS

Problem 48

Three products are produced on four machines (A, B, C, D) where each machine requires a 3.5 m $\times$ 3.5 m area. The routing for each product is given in the following table. Which block layout minimizes the overall flow times distance? (Assume centroid-to-centroid distances.)

	route				loads/day
product 1	A	B	D	C	30
product 2	B	C	D	A	20
product 3	C	A	B	D	40

(A)

A	B
C	D

(B)

A	B	C	D

(C)

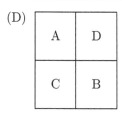

C	D	B	A

(D)

A	D
C	B

Solution

The following is the upper triangular flow matrix (e.g., flow C-D of 50 = flow from C to D [product 2 = 20] and flow from D to C [product 1 = 30]).

from/to	A	B	C	D
A	–	30 + 40 = 70	40	20
B	–	–	20	30 + 40 = 70
C	–	–	–	30 + 20 = 50

Calculate the flow times distance ($F \times D$) score for each solution.

$$F \times D_{\text{solution A}} = (70 \times 10) + (40 \times 10)$$
$$+ (20 \times 20) + (20 \times 20)$$
$$+ (70 \times 10) + (50 \times 10)$$
$$= 3100$$

$$F \times D_{\text{solution B}} = (70 \times 10) + (40 \times 20)$$
$$+ (20 \times 30) + (20 \times 10)$$
$$+ (70 \times 20) + (50 \times 10)$$
$$= 4200$$

$$F \times D_{\text{solution C}} = (70 \times 10) + (40 \times 30)$$
$$+ (20 \times 20) + (20 \times 20)$$
$$+ (70 \times 10) + (50 \times 10)$$
$$= 3900$$

$$F \times D_{\text{solution D}} = (70 \times 20) + (40 \times 10)$$
$$+ (20 \times 10) + (20 \times 10)$$
$$+ (70 \times 10) + (50 \times 20)$$
$$= 3900$$

The answer is A.

Problem 49

A test station is to be incorporated into a fabrication department. The station will receive five loads per day from manufacturing center MC1 at a location given by (x, y) coordinates $(10, 8)$, four from MC2 at $(2, 4)$, three from MC3 at $(4, 6)$, and two from MC4 at $(6, 10)$. Where should the station be located to minimize total distance traveled?

(A) $(6, 6)$
(B) $(4, 6)$
(C) $(6, 4)$
(D) $(4, 8)$

Solution

The optimal value of the test station can be found by solving for each of the two coordinates (x and y) separately using two mathematical properties of the optimal solution.

(1) The x- and y-coordinates of the new facility will be one of the x- and y-coordinates of the existing facilities.

(2) The optimal x and y will be located so that no more than half of the total weight is to the left of x and no more than half is to the right of x.

The application of these two properties gives the following.

$$\tfrac{1}{2} \text{ sum (or total weight)} = \frac{5 + 4 + 3 + 2}{2} = 7$$

	x-coordinate	weight	sum weights
MC2	2	4	4
MC3	4	3	7 = 7
MC4	6	2	9
MC1	10	5	14

The x-coordinate is 4.

	y-coordinate	weight	sum weights
MC2	4	3	3
MC3	6	4	7 = 7
MC1	8	5	12
MC4	10	2	14

The y-coordinate is 6.

The answer is B.

Problem 50

Given the following product demands, how many storage spaces would be required for all four products if a dedicated storage policy was used versus a random storage policy?

period	product 1	product 2	product 3	product 4
1	30	25	5	50
2	40	10	20	30
3	50	15	15	40
4	30	20	30	20
5	20	30	25	25

(A) dedicated = 100, random = 50
(B) dedicated = 120, random = 160
(C) dedicated = 160, random = 100
(D) dedicated = 160, random = 120

Solution

period	product 1	product 2	product 3	product 4	total
1	30	25	5	50	110
2	40	10	20	30	100
3	50	15	15	40	120
4	30	20	30	20	100
5	20	30	25	25	100
maximum	50	30	30	50	120

Dedicated storage space requirements are equal to the sum of the maximum demands for each product.

$$50 + 30 + 30 + 50 = 160$$

Random storage space requirements are equal to the maximum aggregate demand over all time periods.

$$\text{maximum of } (110, 100, 120, 100, 100) = 120$$

The answer is D.

Problem 51

It is most appropriate and economical to use a conveyor-based material handling system when

(A) the material flow rate is medium to high, the distance is short to medium, and the product routing is fixed
(B) the material flow rate is low to medium, the distance varies from short to medium, and the product routing is variable but well defined
(C) the material flow rate is low, the distance is highly variable, and the product routing is highly variable
(D) the material flow rate is high, the distance is long, and the product routing is moderately variable

Solution

Conveyors are most economical when used for high-flow rate moves over short distances. Since the cost to reconfigure a conveyor system is high, conveyors are most appropriate for fixed product routings.

The answer is A.

Problem 52

A recently hired industrial engineer submitted a material handling system design that contains the following processes: (1) transfer, where a gravity conveyor moves a pallet holding a 100 kg, 0.5 m × 0.5 m × 0.5 m box containing four 25 kg gear casings delivered from a long-term contract vendor from the loading dock to incoming inspection 20 m away, (2) incoming inspection, where the cardboard containers for the boxes are discarded and the casings are manually loaded onto an overhead power and free conveyer, and (3) transfer, where the casings are transported in a loop from workstations 1 to 2 to 3 to 4 to 5 to packaging. The system automatically moves parts to the appropriate workstation based on bar code readings. At the packaging station, ten finished 25 kg casings are reboxed and placed on a pallet to be moved by forklift into a waiting truck. Which of the following lists all of the material handling design principles that were violated?

- (A) mechanization, unit load, and simplification
- (B) flexibility, systems, and gravity
- (C) flexibility, ergonomics, and ecology
- (D) ergonomics, gravity, unit load, and systems

Solution

Flexibility was violated because a conveyor system is inherently inflexible and hard to reconfigure if future casing designs require more complex routings. Ergonomics was violated because the loading of a 25 kg casing is beyond the lifting limits for repetitive work. Ecology was violated because the vendor has an established relationship with the company, and the industrial engineer should have worked with the vendor to design a reusable container that would minimize both the cost and waste from transporting the casings.

The answer is C.

Problem 53

A factory with four departments has to determine the type of material handling equipment to use. The departments have determined that either a forklift, a pallet jack, or a conveyor system would be technically feasible. However, the combined capital and operating cost ($ per meter traveled) and the size of the unit load

depend on the type of material handling equipment selected. The following material flow routing, distance, and material handling equipment data are given. What is the total cost associated with each type of equipment, and which should be selected?

material handling type	units/load	$/m
conveyor	100	0.2
forklift	200	1
pallet jack	20	0.05

	routing				no. of units
product 1	D1	D2	D3	D4	1000
product 2	D1	D3	D2	D4	2000

distance (m)	D1	D2	D3	D4
D1	–	50	100	150
D2		–	50	100
D3			–	50
D4				–

- (A) conveyor = $2000, forklift = $2200, pallet jack = $2500; select the conveyor
- (B) conveyor = $1300, forklift = $3250, pallet jack = $1625; select the forklift
- (C) conveyor = $900, forklift = $1550, pallet jack = $550; select the pallet jack
- (D) conveyor = $1300, forklift = $3250, pallet jack = $1625; select the conveyor

Solution

The total distance for product 1 is

$$50 \text{ m} + 50 \text{ m} + 50 \text{ m} = 150 \text{ m}$$

The total distance for product 2 is

$$100 \text{ m} + 50 \text{ m} + 100 \text{ m} = 250 \text{ m}$$

For the conveyor, the total cost is

$$\left(0.2 \ \frac{\$}{\text{m}}\right)\left(\left(\frac{1000 \text{ units}}{100 \ \frac{\text{units}}{\text{load}}}\right)(150 \text{ m}) + \left(\frac{2000 \text{ units}}{100 \ \frac{\text{units}}{\text{load}}}\right)(250 \text{ m})\right) = \$1300$$

For the forklift, the total cost is

$$\left(1\ \frac{\$}{m}\right)\left(\begin{array}{c}\left(\dfrac{1000\ \text{units}}{200\ \dfrac{\text{units}}{\text{load}}}\right)(150\ m) \\ + \left(\dfrac{2000\ \text{units}}{200\ \dfrac{\text{units}}{\text{load}}}\right)(250\ m)\end{array}\right) = \$3250$$

For the pallet jack, the total cost is

$$\left(0.05\ \frac{\$}{m}\right)\left(\begin{array}{c}\left(\dfrac{1000\ \text{units}}{20\ \dfrac{\text{units}}{\text{load}}}\right)(150\ m) \\ + \left(\dfrac{2000\ \text{units}}{20\ \dfrac{\text{units}}{\text{load}}}\right)(250\ m)\end{array}\right) = \$1625$$

The answer is D.

Problem 54

A company is designing a new room for one of its departments. The room will house lathes ($j = 1$) and mills ($j = 2$), two different types of machines. Each lathe will require 10 m^2 of space, and each mill will require 20 m^2 of space. Both the lathes and the mills will be used to process three different types of products ($i = 1, 2,$ and 3). Assuming the department works a 40 hr week, and using the following relevant production data, the area needed to house the machines is most nearly

	weekly production rate		production time (hr)	
	lathes	mills	lathes	mills
product 1	100		0.5	
product 2	200	200	1.0	0.5
product 3		250		1.0

(A) 213 m^2
(B) 220 m^2
(C) 240 m^2
(D) 250 m^2

Solution

To calculate the amount of total space needed to house the lathes and mills, first determine how many of each machine type are required using the following equation.

$$M_j = \sum_{i=1}^{n} \frac{P_{ij}T_{ij}}{C_{ij}}$$

M_j = number of machines of type j required per production period
j = type of machines = lathes; mills
i = types of products
n = number of products = 3
P_{ij} = weekly production rate for product i on machine j
T_{ij} = production time for product i on machine j
C_{ij} = number of hours in production period for product i on machine j

$$M_{\text{lathe}} = \frac{(100)(0.5\ \text{hr}) + (200)(1.0\ \text{hr})}{40\ \text{hr}}$$
$$= 6.25\ \text{lathes}\quad(7\ \text{lathes})$$

$$M_{\text{mill}} = \frac{(200)(0.5\ \text{hr}) + (250)(1.0\ \text{hr})}{40\ \text{hr}}$$
$$= 8.75\ \text{mills}\quad(9\ \text{mills})$$

The combined total space required is

$$(7\ \text{lathes})(10\ m^2) + (9\ \text{mills})(20\ m^2) = 250\ m^2$$

The answer is D.

HUMAN FACTORS, PRODUCTIVITY, ERGONOMICS, AND WORK DESIGN

Problem 55

An industrial engineer is in charge of determining a new product's standard labor cost. From experience, the engineer knows this type of product has a 90% learning curve. The standard time used to compute the standard cost is taken to be the time to produce the 100th unit. The time to produce the first unit is 10 hr. The time required to produce the 100th unit is most nearly

(A) 1.5 hr
(B) 5.0 hr
(C) 8.1 hr
(D) 9.0 hr

Solution

The following equation is used to determine the time required to produce the 100th unit.

$$Z_u = Ku^{(\log s/\log 2)}$$
u = output unit number = 100
Z_u = time to produce output unit number u
K = time to produce the first output unit = 10 hr
s = learning curve slope parameter expressed as a decimal = 0.90

$$Z_{100} = (10\ \text{hr})(100)^{(\log 0.90/\log 2)}$$
$$= 4.966\ \text{hr}\quad(5.0\ \text{hr})$$

The answer is B.

Problem 56

An alarm produces a 3000 Hz sound with a sound level of 70 dB, measured on the A scale and based upon a formula for a person with normal hearing. The area in the plant where this alarm is to be installed has both male and female employees ages 25 through 65. To ensure that all employees can hear this alarm, the sound level should at least be increased to approximately how many decibels measured on the A scale?

- (A) 75 dBA
- (B) 90 dBA
- (C) 100 dBA
- (D) 120 dBA

Solution

The answer must address the average shifts in age of the threshold of hearing for pure tones of persons with normal hearing. Using the graphs in the NCEES Handbook and designing for the worst condition at age 65, the shift for men at age 65 is 30 dBA, and the shift for women at age 65 is 22 dBA.

Add 30 dBA to the current sound level of the alarm.

$$70 \text{ dBA} + 30 \text{ dBA} = 100 \text{ dBA}$$

The answer is C.

Problem 57

An adjustable seat is being designed to accommodate both males and females. Assume shoe height is 3 cm and the design needs to adjust to accommodate 5% to 95% of both males and females. The adjustment range in centimeters should be about

- (A) 39–52 cm
- (B) 35–55 cm
- (C) 39–42 cm
- (D) 37–51 cm

Solution

Use the ergonomic table in the NCEES Handbook.

Seating height plus 3 cm for shoes is as follows.

	percentiles	
	5th	95th
female	38.5	47.3
male	42.2	51.8

Use 38.5–51.8 cm (39–52 cm).

The answer is A.

Problem 58

A company has selected a standard chair height of 48 cm and a standard table height of 74 cm. Any tables under consideration must have a 3 cm thick top. What would be the maximum thickness of the support for the table top if 95% of all employees must be able to put their legs under the table? Assume shoe thickness is 3 cm.

- (A) 3.5 cm
- (B) 5.3 cm
- (C) 6.8 cm
- (D) 10 cm

Solution

Look at the thigh thickness 95th percentile.

$$3 \text{ cm} + h + T + 48 \text{ cm} = 74 \text{ cm}$$

$$48 \text{ cm} = \text{height of chair}$$
$$74 \text{ cm} = \text{height of table}$$
$$3 \text{ cm} = \text{thickness of table}$$
$$h = \text{thickness of support in cm}$$
$$T = \text{thigh clearance height in cm}$$

From the ergonomics table in the NCEES Handbook, the thigh thickness (T) 95th percentile is 17.7 cm for men, and 17.5 cm for women.

$$T = 17.7 \text{ cm} \left(\begin{array}{c} \text{the larger of the values} \\ \text{for men and women} \end{array} \right)$$

$$3 \text{ cm} + h + 17.7 \text{ cm} + 48 \text{ cm} = 74 \text{ cm}$$

$$h = 5.3 \text{ cm}$$

The answer is B.

Problem 59

Consider the following work sampling study for a piece of production equipment based upon a random sample of 100 observations.

activity	no. of times observed
machine idle—no product	5
machine operating	85
machine idle—downtime and repair	10

Find the 95% confidence interval for the proportion of the time the machine is idle.

- (A) [0.05, 0.25]
- (B) [0.08, 0.22]
- (C) [0.10, 0.20]
- (D) [0.14, 0.16]

Solution

The general formula for finding the 95% confidence interval for a proportion that an activity occurs is

$$p \pm z\sqrt{\frac{p(1-p)}{n}}$$

n = total number of observations

x = times the activity is observed

$p = \dfrac{x}{n}$

$z = 1.96$ = standardized normal variant for a 95% confidence interval allocating $2\frac{1}{2}$% to each tail (i.e., a 2-tail analysis since the confidence limits are for both upper and lower bounds)

For this problem,

$$x = 5 + 10 = 15$$
$$n = 100$$
$$p = \frac{15}{100} = 0.15$$

The confidence interval is

$$0.15 \pm 1.96\sqrt{\frac{(0.15)(1-0.15)}{100}} = 0.15 \pm 0.07$$
$$= [0.08, 0.22]$$

The answer is B.

Problem 60

The allowed production time from a time study is 2.3 min per unit. Assume workers are allowed two 15 min breaks for personal allowances. Combined allowances for fatigue and unavoidable delay are 15%. How many parts should a worker most nearly produce during an 8 hr shift?

(A) 50 parts/day
(B) 100 parts/day
(C) 170 parts/day
(D) 200 parts/day

Solution

The standard time is

(allowed time)(1 + allowances as a decimal)

$$= \left(2.3 \ \frac{\text{min}}{\text{unit}}\right)(1 + 0.15)$$
$$= 2.645 \ \text{min/unit}$$

For an 8 hr shift with two 15 min breaks, the production time is

$$\left(8 \ \frac{\text{hr}}{\text{day}}\right)\left(60 \ \frac{\text{min}}{\text{hr}}\right) - (2)\left(15 \ \frac{\text{min}}{\text{day}}\right)$$
$$= 450 \ \text{min/day}$$

The number of parts per day a worker should produce is

$$\left(450 \ \frac{\text{min}}{\text{day}}\right)\left(\frac{1 \ \text{part}}{2.645 \ \text{min}}\right)$$
$$= 170.13 \ \text{parts/day} \quad (170 \ \text{parts/day})$$

The answer is C.

Problem 61

A coil-slitting operation consists of three elements. A summary of a time study and a work sampling study is shown with all times in minutes per unit.

element	cycles timed	average time	occurrences per cycle	rating
1	35	0.64	1	100
2	35	0.55	1	110
3	35	1.28	1	95

work sampling study

activity	no. of times observed
machine idle—no material available	20
machine idle—maintenance	10
machine working	150
operator away from machine—avoidable	5
operator away from machine—personal	15
	200

Assume there are no official breaks for the employee. The standard time in minutes per unit based upon the results of this study would most nearly be

(A) 2.5
(B) 2.8
(C) 2.9
(D) 3.0

Solution

Use the following equation to solve for the standard time (ST).

$$ST = AT(1 + A)$$
$$A = \text{allowances}$$

OT_i = observed average time for element i
n = number of elements
R_i = rating factor

$$AT = \text{allowed time} = \sum_{i=1}^{n} OT_i \left(\frac{R_1}{100}\right)$$

$$= \left(0.64 \ \frac{\text{min}}{\text{unit}}\right)(1.00) + \left(0.55 \ \frac{\text{min}}{\text{unit}}\right)(1.10)$$

$$+ \left(1.28 \ \frac{\text{min}}{\text{unit}}\right)(0.95)$$

$$= 2.461 \ \text{min}$$

Estimate allowances from the work sampling study.

$$\text{allowance for idle machine} = \frac{20+10}{200} = 0.15$$

$$\text{personal allowance} = \frac{15}{200} = 0.075$$

$$A = 0.15 + 0.075 = 0.225$$

$$ST = \left(2.461 \ \frac{\text{min}}{\text{unit}}\right)(1.225)$$

$$= 3.0147 \ \text{min/unit} \quad (3.0 \ \text{min/unit})$$

The answer is D.

Problem 62

A cart designed for pushing is 159.6 cm in overall height. Assuming that height is normally distributed, approximately what percent of women pushing the cart will be able to see over the top? Assume that shoe height is 3 cm.

(A) 5%
(B) 8%
(C) 12%
(D) 21%

Solution

To find the percent of women who will be able to see over the top of the cart, first add the mean women's eye height to the given shoe height to find the eye height from the floor.

$$x = \text{eye height from floor}$$

From the NCEES Handbook's ergonomics table, the barefoot mean eye height is 148.9 cm.

$$x = 148.9 \ \text{cm} + 3 \ \text{cm} = 151.9 \ \text{cm}$$

The standard deviation is 6.4 cm.

$$P(x > 159.6 \ \text{cm}) = P\left(z > \frac{159.6 \ \text{cm} - 151.9 \ \text{cm}}{6.4 \ \text{cm}}\right)$$

$$= P(z > 1.20)$$

$$= 0.1151 \quad (12\%)$$

The answer is C.

Problem 63

An assembly department has been complaining that it cannot meet its production standard due to the amount of time it has to spend finding missing parts. To determine the percentage of time that an assembly department spends looking for missing parts, an industrial engineer decides to conduct a work sampling study. If the department has been reporting that 20% of its time is used looking for missing parts, approximately how many observations must the industrial engineer make in order to prove the department's claim that it is 90% certain that the maximum error of its estimate is ±5%?

(A) 40
(B) 120
(C) 1700
(D) 4300

Solution

Use the following equation to find the number of observations.

$$n = \frac{p(1-p)}{\left(\dfrac{Sp}{z}\right)^2}$$

n = number of observations
p = percent of time an activity occurs = 20%
S = degree of accuracy required = ±5%
   Since the estimate may be either 5% too high or too low, this is a 2-tail analysis. Allocate 100% − 90%/2 = 5% to each tail.
z = standard normal distribution variant for 90% confidence = 1.64

$$n = \frac{(0.2)(1-0.2)}{\left(\dfrac{(0.05)(0.2)}{1.64}\right)^2}$$

$$= 4303 \ \text{observations} \quad (4300 \ \text{observations})$$

The answer is D.

Problem 64

The following symbols are used to draw a flow process chart.

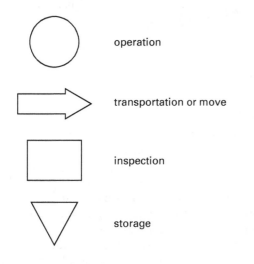

operation

transportation or move

inspection

storage

In preparation of a flow process chart, which of the following tasks would NOT be considered an operation?

(A) driving a nail into a piece of wood
(B) drilling a hole in a casting using a drill press
(C) loading a part into a drill fixture
(D) examining a paint finish for scratches

Solution

Examining a paint finish for scratches would be considered an inspection, not an operation.

The answer is D.

QUALITY

Problem 65

A company uses a single-attribute sampling plan on incoming critical materials.

Three parts are drawn at random from a carload of 50,000 parts and are subjected to a stress yield test. If all three pieces meet the minimum yield strength, the entire lot is accepted. If one or more parts fail the test, the entire lot is returned to the supplier.

If the lot is submitted from a process whose population mean proportion defective rate is 10%, and when approximated with a Poisson distribution, the probability as a percentage of accepting this lot is most nearly

(A) 0%
(B) 10%
(C) 40%
(D) 75%

Solution

Use the following problem parameters and the Poisson distribution with $x = c$ to approximate the binomial probabilities.

$$N = \text{lot size} = 50{,}000$$
$$n = \text{random sample size} = 3$$
$$c = \text{acceptance number} = 0$$

$$P(x = 0) = \frac{e^{-nP}(nP)^x}{x!} = \frac{e^{-(3)(0.10)}\big((3)(0.10)\big)^0}{0!}$$
$$= 0.7408 \quad (75\%)$$

The answer is D.

Problem 66

The specification limits for an axle turned on a lathe had the following blueprint specifications for the diameter. (Note that it is possible to adjust the lathes, thus moving the population mean.)

$$285 \pm 6.0$$

A trial control chart on the process yielded the following limits for $\overline{X}$ and R based on 23 samples of size 5.

$$\text{UCL}_{\overline{X}} = 294.16$$
$$\text{LCL}_{\overline{X}} = 283.76$$
$$\text{UCL}_R = 28.910$$
$$\text{LCL}_R = 0$$

All sample points fell within the trial control limits $(K = 23)$.

$$\overline{\overline{X}} = \frac{\sum \overline{X}_i}{K} = 288.96$$
$$\overline{R} = \frac{\sum R_i}{K} = 8.960$$

Assuming that the population from which the data were drawn is approximately normally distributed, what is the population mean?

(A) 285.00
(B) 288.90
(C) 288.96
(D) 289.06

Solution

$$\mu_x \approx \overline{\overline{X}} = 288.96$$

The answer is C.

Problem 67

Using the information given in Prob. 66, and assuming permanent $\overline{X}$ and R control charts are needed for this process, the $\text{UCL}_{\overline{X}}$ and $\text{LCL}_{\overline{X}}$ limits should be most nearly set at

(A) 290.2; 279.8
(B) 291.0; 279.0
(C) 292.2; 281.1
(D) 294.2; 283.8

Solution

The information given in Prob. 66 states that the lathe cutting diameter can be adjusted, so adjust the population mean to the desired setting ($\mu_0 = 285.0$, the specified diameter).

$$\sigma_{\overline{X}} \approx \frac{\overline{R}}{d_{2,n=5}} = \frac{8.960}{2.326} = 3.852$$

The permanent control chart limits for $\overline{X}$ would be $\mu_0 \pm 3(\sigma_{\overline{X}}/\sqrt{n})$ where $\sigma_{\overline{X}} = 3.852$ and $n = 5$.

$$\text{UCL}_{\overline{X}} = 285.0 + (3)\left(\frac{3.852}{\sqrt{5}}\right)$$
$$= 290.17 \quad (290.2)$$

$$\text{LCL}_{\overline{X}} = 285.0 - (3)\left(\frac{3.852}{\sqrt{5}}\right)$$
$$= 279.83 \quad (279.8)$$

The answer is A.

Problem 68

The ISO 9000 is best described as

(A) a top-to-bottom quality management system
(B) a system limited to foreign exporters and importers
(C) an easily implemented quality management system
(D) a quality management system applicable only to firms of 500 or more employees

Solution

ISO 9000 is a group of top-to-bottom quality management systems that give an organization a good start on implementing total quality.

The answer is A.

Problem 69

"Empowerment" in total quality management (TQM) teams essentially means the team

(A) is given all necessary resources to solve the assigned problem
(B) is limited by self-imposed constraints
(C) is limited by management imposed constraints
(D) must report only to the immediate supervisor of the problem area

Solution

In TQM teams, empowerment means the team is limited by management imposed constraints. Goetoch and Davis, in their book *Introduction to Total Quality*, state one way this can be done is by structuring work that allows employees to make decisions concerning the improvement of work processes within well-defined parameters. These parameters represent constraints of solutions within employees' spheres of knowledge.

The answer is C.

Problem 70

W. Edward Deming is most noted for his

(A) development of the 14 points an organization should follow to achieve total quality
(B) contribution to the development of the Malcomb-Baldrige Award
(C) controversial "point" of eliminating quotas
(D) great success in implementing TQM in the U.S. steel industry

Solution

W. Edward Deming is most noted for his 14 points that describe what is necessary for a business to survive, be competitive, and achieve total quality.

The answer is A.

Problem 71

When setting up a quality system, it is necessary to determine the aspects of product quality that need to be considered. Which of the following are NOT considered dimensions of quality?

(A) performance, features
(B) reliability, conformance
(C) durability, serviceability
(D) flexibility, cost

Solution

David A. Garvin, in his 1984 article "What Does 'Product Quality' Really Mean?" identified eight dimensions of quality that capture the basic elements of a quality product. These dimensions include performance, features, reliability, conformance, durability, serviceability, aesthetics, and perceived quality. Flexibility and cost are not considered dimensions of quality.

The answer is D.

Practice Exam 1

PROBLEMS

1. Jane is planning for her retirement. Each month she places $200 in an account that pays 12% nominal interest, compounded monthly. She made the first deposit of $200 on January 31, 1997. The last $200 deposit will be made on December 31, 2016. If the interest rate remains constant and all deposits are made as planned, the amount in Jane's retirement account on January 1, 2017 is most nearly

(A) $60,000
(B) $155,000
(C) $173,000
(D) $198,000

2. Dawn purchased a $10,000 car. She put $2000 down and financed the $8000 balance. The interest rate is 9% nominal, compounded monthly, and the loan is to be repaid in equal monthly installments over the next four years. Dawn's monthly car payment is most nearly

(A) $167
(B) $172
(C) $188
(D) $200

3. James is a major prizewinner in a sweepstakes. He has the option of either receiving a single check for $125,000 now or receiving a check for $50,000 each year for three years. (James would be given the first $50,000 check now.) At what interest rate would James most nearly have to invest his winnings for him to be indifferent as to how he receives his winnings?

(A) 16%
(B) 20%
(C) 22%
(D) 23%

4. In the design of an automobile radiator, an engineer has a choice of using either a brass-copper alloy casting or a plastic molding. Either material provides the same service. However, the brass-copper alloy casting has a mass of 25 lbm, compared with 16 lbm for the plastic molding. Every pound of extra mass in the automobile has been assigned a penalty of $4 to account for increased fuel consumption during the lifecycle of the car. The brass-copper alloy casting costs $3.35 per pound,

whereas the plastic molding costs $7.40 per pound. Machining costs per casting are $6.00 for the brass-copper alloy. Which material should the engineer select, and what is the difference in unit cost?

(A) brass-copper alloy, savings = $1.35/radiator
(B) brass-copper alloy, savings = $6.00/radiator
(C) plastic, savings = $7.35/radiator
(D) plastic, savings = $8.50/radiator

5. A process engineer is trying to decide on the type of tool material to use for a machining operation. Relevant data for the alternatives are as follows.

	tool material A	tool material B
tool cost	$100	$30
production rate	100 pieces/hr	75 pieces/hr
tool life	50 hr	25 hr

It takes 1 hour to replace a worn tool. Labor costs $18 per hour, and a total of 15,000 pieces are needed. What tool material should be selected, and what are the expected savings?

(A) material A, save $900
(B) material B, save $1000
(C) material A, save $1000
(D) material B, save $3100

6. A project engineer is deciding on the most cost-effective duration for a new project. The direct costs of the project are expected to vary indirectly with project duration, while indirect costs vary directly with the square of the project duration. The total project cost is represented by the following equation.

$$\text{project cost} = \$5000 + \frac{\$4000 \text{ mo}}{T} + \frac{\$250 T^2}{\text{mo}^2}$$

$$T = \text{project duration in months}$$

For what duration should the project be planned, and what is most nearly the expected project cost?

(A) 3 mo, $600
(B) 2 mo, $3000
(C) 2 mo, $8000
(D) 3 mo, $8600

7. The state of Colorado is considering opening a new state park in the Rocky Mountains. Park development costs are estimated to be $20 million, and operating expenses are estimated to be $2 million per year. It is anticipated that 500,000 people per year will visit the park, and the park entry fee will be $3 per person. State economists have calculated a positive economic impact on the area to be $15 per person per visit due to lodging, food, souvenirs, and other factors. Assuming a 6% annual interest rate and a 20 year planning horizon, the benefit-cost (B/C) ratio for this project is most nearly

    (A) 0.4
    (B) 0.9
    (C) 2.4
    (D) 3.0

8. A machine is purchased for $80,000 and has annual operating expenses of $10,000. What is the pay-back period on the machine if it generates revenues of $25,000 per year?

    (A) 5 yr
    (B) 6 yr
    (C) 7 yr
    (D) 8 yr

9. The following cash flow profiles are for two mutually exclusive alternatives, A_1 and A_2, of 1 year duration each.

t	A_1	A_2
0	−$500	−$1000
1	+$560	+$1110

Alternative A_1 has already been determined feasible. What is the incremental internal rate of return of alternative A_2 compared to A_1? Use a minimum attractive rate of return of 9%.

    (A) 10%
    (B) 11%
    (C) 12%
    (D) 13%

10. A two-way factorial design is used to determine whether a treatment is significant at a 0.05 level. The analysis of a variance table from experimental results is as follows.

source of variation	degrees of freedom	sum of squares
replications	4	122.46
treatments	4	126.92
residuals	16	116.08

Which of the following statements is true?

    (A) One can conclude that the treatment is significant at a 0.05 level.
    (B) One can conclude that the treatment is not significant at a 0.05 level.
    (C) The replication error is significant, thus one cannot make a conclusion.
    (D) The replication error is significant, thus one can conclude that the treatment is significant at a 0.05 level.

11. A 2^2 factorial design with three replications is used to determine significant factors at an 0.05 level. The analysis of the variance table from experimental results is as follows.

source of variation	degrees of freedom	sum of squares
factor A	1	108.50
factor B	1	61.22
cross term AB	1	9.60
errors	8	32.00

One can conclude that

    (A) factor A is significant at an 0.05 level
    (B) factor A and B are both significant at an 0.05 level
    (C) factor A and B and cross term AB are all significant at an 0.05 level
    (D) none of the factors are significant

12. A 3^k factorial design is used to determine significant factors. However, there are no replications for any combination during experiments. Without conducting any more experiments, the best one can do is

    (A) nothing, since there are no error terms
    (B) interpolate error terms and reduce one degree of freedom, then conduct the analysis of variance as usual
    (C) pool the high-order cross terms as error terms, then conduct the analysis of variance
    (D) use the most significant factor as error terms, then conduct the analysis of variance

13. The service time (in hours) for a copy machine is approximately exponentially distributed. By examining the records for 50 breakdowns, it is determined that the average service time is 1.25 hr. The probability that a service time will exceed 2 hr is approximately

    (A) 0.05
    (B) 0.10
    (C) 0.15
    (D) 0.20

14. An engineer would like to test the null hypothesis at a 5% level of significance that the mean shear strength of spot welds is at least 450 psi. The engineer randomly selects 15 welds, measures the shear strength, and determines the sample mean ($\bar{x}$) is 445 psi, and the sample standard deviation (s) is 10 psi.

Based upon the data,

    (A) the null hypothesis is true
    (B) the null hypothesis is false
    (C) there is not enough information to say the hypothesis is true
    (D) there is not enough information to say the hypothesis is false

15. The p-value is the smallest level of significance that would lead to rejection of the null hypothesis. Using the information given in Prob. 14, the p-value is approximately

    (A) 0.001
    (B) 0.005
    (C) 0.025
    (D) 0.040

16. A simulation model was run for 30 replications and the mean utilization of a transporter was recorded for each replication. Those 30 data points were then used to form a confidence interval on mean transporter utilization for the system. At a 95% confidence level, that interval was found to be 37.2% ± 3.4%.

Given this information, which of the following facts can be definitively stated about the system?

    (A) At 95% confidence, the 30-replication sample mean of transporter utilization lies in the range 37.2% ± 3.4%.
    (B) At 95% confidence, the population mean of transporter utilization lies in the range 37.2% ± 3.4%.
    (C) At 95% confidence, the transporter never reached 100% utilization at any point during any of the 30 replications.
    (D) With only 30 replications, no definitive statements can be made about the system, including the utilization of the transporter.

17. If a lot of size $n = 3$ is submitted from a process that has a population mean proportion defective of 30%, the probability of accepting this lot is most nearly

    (A) 0%
    (B) 10%
    (C) 40%
    (D) 75%

18. A manufacturer notices that during the packaging process, 10% of the products become damaged. Out of a sample of 10 products, the probability that two products will be damaged is most nearly

    (A) 0.10
    (B) 0.19
    (C) 0.22
    (D) 0.36

19. A spreadsheet has been developed to calculate the cycle time for an assembly line required to meet a given production demand. Cells A1 through A5 contain the task times, cell B1 contains the number of stations, and cell B2 contains the cycle time. If cell C1 is to contain the balance delay for the solution, what formula should be put into it?

    (A) (SUM(A1...A5) + B1*B2)/(B1 + B2)
    (B) (B1 − (SUM(A1...A5)/B1*B2))/ (SUM(B1...B2))
    (C) (B1*B2 − SUM(A1...A5))/SUM(A1...A5)
    (D) (B1*B2 − SUM(A1...A5))/B1*B2

20. A data file contains the last 12 months of sales data from the Ohio Valley region. The manager is trying to determine the number of sales personnel to assign to the region for the next month. She has written a pseudocode segment to forecast the expected sales for next month and to assign the number of personnel. The data are

$$400, 500, 600, 400, 800, 850, 950,$$
$$1100, 1150, 1300, 1320, 1440$$

The pseudocode segment is as follows.

```
N = 1
Read data point
While N < 8
Increment N by 1
Read data point
Endwhile
Initialize F to 0 and set N = 1
While N < 6
Increment N by 1
Read data point and add value to F
Endwhile
Set P = F/(N − 1)
If P < 1000 set SF = 5
If P > 999 and P < 1300 set SF = 8
If P > 1299 and P < 1450 set SF = 10
Else set SF = 14
```

The number of sales personnel, SF, at the end of this segment would be

(A) 5
(B) 8
(C) 10
(D) 14

21. Consider the problem of assigning four operators to four machines. The assignment costs in dollars are given in the following matrix. Operator 1 cannot be assigned to machine 2, and operator 4 cannot be assigned to machine 4. What is the minimum cost assignment for this problem?

	machine			
	1	2	3	4
operator 1	20	–	20	8
operator 2	14	16	8	12
operator 3	36	12	20	28
operator 4	28	8	24	–

(A) $36
(B) $48
(C) $50
(D) $68

22. Consider a transportation problem that has the following supplies, demands, and cost matrix C where the rows represent the sources and the columns represent the destinations.

source	1	2	3	4	
supply	20	40	10	60	

destination	1	2	3	4	5
demand	5	15	40	35	35

C	destination				
	10	15	5	3	12
	11	4	12	23	19
source	9	13	24	28	21
	23	34	42	51	29

The minimal cost solution generated by the northwest corner rule would be most nearly

(A) 85
(B) 130
(C) 2600
(D) 3300

23. Consider two stocks. Stock 1 always sells for either $20 or $40. If stock 1 is selling for $20 today, there is a 70% chance that it will sell for $20 tomorrow. If it is selling for $40 today, there is an 85% chance it will sell for $40 tomorrow. Stock 2 always sells for either $20 or $35. If stock 2 is selling for $20 today, there is a 90% chance that it will sell for $20 tomorrow. If it is selling for $35 today there is an 80% chance it will sell for $35 tomorrow. What is the expected value of stock 2?

(A) $20
(B) $25
(C) $28
(D) $46

24. The final machining of a product requires three different stages. The stages of the machining are consecutive. Each stage requires a single machine doing a specific task. The machining time at each stage is exponentially distributed with a mean of 20 min. A product must wait to begin the machining process at stage 1 until the current product has completed stage 3. Products arrive at stage 1 according to a Poisson process at the rate of two per hour, and the facility has a buffer that can hold at most ten of these products as they wait for machining. Using Kendall's notation, what type of queuing system is this?

(A) $M/M/3/10$
(B) $M/M/10/3$
(C) $M/G/1/3/10$
(D) $M/E_3/1/11$

25. A university's food court has two tellers. People waiting to pay for their food form a single line and go to the first available teller. There is room in the food court for 200 people, and there is also room outside to wait. Customers have an interarrival time to the food court that is exponentially distributed with a mean of 2 min. The two tellers work at the same rate. The service time for each teller is exponentially distributed with a mean of 3 min. On average, the number of people waiting in line for the tellers is most nearly

(A) 0
(B) 1
(C) 2
(D) 30

26. A company's performance appraisal system is based on each employee establishing, in conjunction with a supervisor, a set of specific objectives against which the employee will be evaluated at a predetermined interval. Which of the following best represents this type of appraisal system?

(A) value based management
(B) performance management
(C) change management
(D) management by objectives

27. A company is considering implementing a 50-50 gain sharing plan, where a base rate of $12.00 per hour is guaranteed, and an employee will receive 50% of the gain after exceeding 100% of the standard. If the standard time for a task is 2.5 minutes, what would a worker get paid by averaging 30 units per hour for an eight-hour day?

(A) $75
(B) $96
(C) $100
(D) $108

Problems 28 and 29 refer to the following project network and associated activity durations.

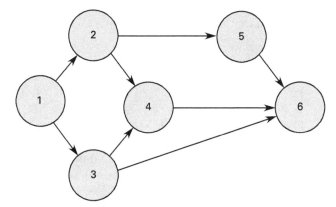

activity	duration (hr)
1-2	4
1-3	3
2-4	6
2-5	10
3-4	2
3-6	16
4-6	8
5-6	2

28. What is the critical path, or longest duration activity sequence, for the project network?

(A) {1-3, 3-6}
(B) {1-2, 2-5, 5-6}
(C) {1-2, 2-4, 4-6}
(D) {1-3, 3-4, 4-6}

29. What is the early start schedule for the project network where A_{i-j} denotes the earliest start for activity i-j?

(A) $A_{1-2} = 0$, $A_{1-3} = 0$, $A_{2-4} = 4$, $A_{2-5} = 4$, $A_{3-4} = 3$, $A_{3-6} = 3$, $A_{4-6} = 10$, $A_{5-6} = 14$
(B) $A_{1-2} = 4$, $A_{1-3} = 3$, $A_{2-4} = 6$, $A_{2-5} = 10$, $A_{3-4} = 2$, $A_{3-6} = 16$, $A_{4-6} = 8$, $A_{5-6} = 2$
(C) $A_{1-2} = 0$, $A_{1-3} = 0$, $A_{2-4} = 4$, $A_{2-5} = 4$, $A_{3-4} = 3$, $A_{3-6} = 3$, $A_{4-6} = 5$, $A_{5-6} = 14$
(D) $A_{1-2} = 4$, $A_{1-3} = 16$, $A_{2-4} = 7$, $A_{2-5} = 17$, $A_{3-4} = 9$, $A_{3-6} = 3$, $A_{4-6} = 11$, $A_{5-6} = 19$

30. A company president is excited about implementing a new approach to motivate employees. The approach's main characteristics are 1) employees will have lifetime employment, 2) promotion will be based on length of service, 3) there will be no employee job specialization, and 4) the company will use a collective decision making process. Which motivation theory does the president's new approach most closely describe?

(A) Theory W
(B) Theory X
(C) Theory Y
(D) Theory Z

31. A project has a critical path with mean $\mu = 20$ and standard deviation $\sigma = 2$, and a non-critical path with mean $\mu = 18.5$ and standard deviation $\sigma = 6$. Each path is independent. What is most nearly the probability of the project being completed in 21 days?

(A) 0.42
(B) 0.50
(C) 0.66
(D) 0.69

32. A part requires that two operations (Op1 followed by Op2) be processed on a single machine. The company will be operating 5 days per week and 10 hours per day with no set-up time between operations. The following data is available for the operations.

operation	Op1	Op2
standard time (in minutes)	2.0	4.0
actual performance	0.90	0.90
machine availability	0.90	0.95
scrap rate	0.03	0.05

What is the number of machines required to produce 5000 parts per week?

(A) 10
(B) 12
(C) 13
(D) 15

33. A single-point tool is used on an engine lathe to make a single pass on a part 2.000 in diameter (diameter before the cut), and 24 in long. The depth of cut is set to 0.125 in at a spindle speed of 400 rev/min with a feed of 0.012 in/rev.

The machining time required for a single pass is most nearly

- (A) 0.21 min
- (B) 5.0 min
- (C) 10 min
- (D) 115 min

34. A short-stroke reciprocating saw can cut ductile steel at about 12 in^2/min. The average time for the operator to unclamp, clamp, and advance stock in the saw is 15 sec per cut. The saw is currently loaded with 3 in diameter, low-carbon, steel bar stock.

Approximately how long will it take to cut twelve 4 in pieces?

- (A) 0.21 min
- (B) 5.0 min
- (C) 10 min
- (D) 120 min

35. The wheel assembly for a rolling cart requires 11 components (wheels, nuts, bolts, washers, and the yoke), two hand tools (a screwdriver and pliers), and two assembly directions (from the side and the bottom). Which of the following guidelines for the manufacturing and design of the assembly would be appropriate for improving this product?

rule 1:   Avoid flexible components that are difficult to handle.

rule 2:   Use the simplest possible operations.

rule 3:   Reduce the number of components.

rule 4:   Exaggerate asymmetry when not possible to provide symmetry.

rule 5:   Use snap fasteners if possible.

- (A) rules 1 and 5
- (B) rules 3 and 5
- (C) rules 1, 2, and 4
- (D) rules 2, 3, and 4

36. Jobs go through a painting station and then an automatic inspection station. Each station can only service one job at a time. Four jobs are currently awaiting processing. The time each job must spend at each station is shown as follows.

job	painting time (min)	inspection time (min)
1	12	9
2	8	5
3	15	8
4	7	4

The makespan of the job sequence $\{3, 2, 1, 4\}$ is

- (A) 41 min
- (B) 44 min
- (C) 46 min
- (D) 48 min

37. Component Y is used in several products assembled by company ABC. The average demand for component Y is 750 units per week. The fixed cost of placing an order for component Y is $15. The inventory holding cost for component Y is $0.10 per unit per week. To minimize the total cost of ordering and inventory holding, component Y would be most nearly economically ordered in quantities of

- (A) 100 units
- (B) 320 units
- (C) 470 units
- (D) 750 units

38. The demand for product Z is 10,000 units per week. Each unit of product Z requires 0.25 labor hours to produce. There are two 8.5 hr production shifts per day, five days per week. During each shift, approximately 1.5 hours are nonproductive (due to breaks, setup, and so forth). The minimum workforce needed to meet the demand for product Z without requiring overtime is

- (A) 35 people/day
- (B) 36 people/day
- (C) 35 people/shift
- (D) 36 people/shift

39. A company is considering adopting a just-in-time approach to production and needs to determine the number of kanban cards required to control the system. One of the company's products has a daily demand of 500 units and a processing time of 1 minute per unit. The lead time for producing the part is 2 days, where 8 production hours per day includes the processing and waiting time and where parts are moved in containers that hold 100 units. Determine the number of kanban cards for this product with a safety factor of 10%.

- (A) 5
- (B) 7
- (C) 10
- (D) 11

40. Six assembly cells (each 35 m × 35 m) are to be located in an area that will configure into a 2 cell × 3 cell facility. The cost of material handling is directly proportional to the distance traveled from centroid to centroid. The following material flow data is given.

cell	C1	C2	C3	C4	C5	C6
C1	–	150		300	0	50
C2		–	50	0	10	80
C3			–	150	100	30
C4				–	40	0
C5					–	100
C6						–

If the current cell layout is shown in the figure below, what would be the impact on the total material handling flow if C2 and C4 were to exchange positions?

C1	C3	C4
C6	C5	C2

- (A) −1200 (decrease)
- (B) 0 (no change)
- (C) 2000 (increase)
- (D) 3600 (increase)

41. Consider a warehouse that contains 40 storage locations. Four products are stored in the warehouse. Storage and throughput requirements are given for each product. In what order (or priority) should products be assigned to storage locations to minimize the total distance traveled?

product	no. of locations required S	total loads moved per day T
1	10	200
2	16	160
3	8	240
4	6	150

- (A) $\{2, 1, 3, 4\}$
- (B) $\{2, 1, 4, 3\}$
- (C) $\{3, 1, 2, 4\}$
- (D) $\{3, 4, 1, 2\}$

42. Consider an open-loop conveyor system design problem where parts arrive at a 1.2 m wide conveyor input point at a rate of 100 parts per minute. Five parts are first put on a 1 m × 1.3 m pallet, then the pallets are transported on the conveyor. The distance from the loading to unloading point is 200 m. Calculate the minimum speed of the conveyor to support the specified material flow.

- (A) 20 m/min
- (B) 22 m/min
- (C) 26 m/min
- (D) 30 m/min

43. A production system requires 40 material-handling moves per hour, with each move having an average distance of 240 m. Each move requires a 1 min load and a 1 min unload. Four forklifts are used to move the loads, and each travels at an average of 80 m/min. The average utilization for the four forklifts is most nearly

- (A) 50%
- (B) 70%
- (C) 85%
- (D) 95%

44. A company produces two products that use a combination of six production processes arranged into three cells. Due to the size of the products, a forklift has been selected as the material handling equipment. What is the intercellular material handling flow associated with the following cellular layout? (Assume material handling cost is proportional to distance.)

	routing						loads
product 1	P1	P2	P3	P4	P5	P6	100
product 2	P2	P5	P4	P3	P1	P6	200

distance between cells in meters (processes in cell)	cell 1 (P1, P2)	cell 2 (P3, P4)	cell 3 (P5, P6)
C1 (P1, P2)	–	50	100
C2 (P3, P4)		–	50
C3 (P5, P6)			–

- (A) 6500
- (B) 28,000
- (C) 51,000
- (D) 70,000

45. Four departments are located in a 30 m × 20 m facility, where each department is 10 m × 10 m. There remain two potential locations to locate a single storage facility (as noted in the following illustration as storage location A and storage location B). The flow (units per yr) between the existing four departments and the future storage department is given in the following table.

	dept. 1	dept. 2	dept. 3	dept. 4
flow (unit per yr) between storage and department	250	400	300	600

The transportation cost is \$1/m. Units travel along a rectilinear path between department centroids. Determine the best storage location (either storage A or storage B), and determine its associated flow times distance cost $(F \times D)$.

(A) $F \times D_{\text{storage A}} = \$15,500$
(B) $F \times D_{\text{storage B}} = \$21,000$
(C) $F \times D_{\text{storage A}} = \$25,500$
(D) $F \times D_{\text{storage B}} = \$46,500$

46. A company is setting up a warehouse and will use a class-based storage policy. The following data have been collected on the 12 products that will be stored in the facility. Determine three product classes based on an ABC inventory breakdown, where A, B, and C products make up approximately 50%, 30%, and 20% of the total warehouse activity.

product	throughput T	space S	T/S
1	3.6	3	1.2
2	10.0	4	2.5
3	5.0	12	0.4
4	26.0	3	8.7
5	6.4	4	1.6
6	1.2	4	0.3
7	8.4	5	1.7
8	27.0	6	4.5
9	7.2	8	0.9
10	5.5	5	1.1
11	56.0	9	6.2
12	7.0	10	0.7
totals	163.3	73	29.8

(A) class A = (4, 11),
    class B = (8, 2, 7),
    class C = (5, 1, 10, 9, 12, 3, 6)
(B) class A = (1, 2, 3, 4, 5, 6),
    class B = (7, 8, 9, 10),
    class C = (11, 12)
(C) class A = (11, 8),
    class B = (4, 2, 7),
    class C = (9, 12, 5, 10, 3, 1, 6)
(D) class A = (4, 11, 8, 2, 7, 5),
    class B = (1, 10, 9, 12),
    class C = (3, 6)

47. A chair with armrests is being designed. The clothing allowance is 1.2 cm. What should be the distance between armrests to accommodate 95% of the population?

(A) 25 cm
(B) 33 cm
(C) 38 cm
(D) 45 cm

48. An operation that occurs throughout an 8 hr shift lifts a casting that is located on a pallet an average of 12 in off the floor and 16 in in front of the operator. The casting is lifted vertically an average total distance of 48 in. This lift occurs once every 2.5 min. The maximum frequency of lifting that can be sustained for this job is 12 lifts/min. Using the NIOSH formula, what is the maximum permissible limit (in pounds per force) for the weight of the casting?

(A) 40 lbf
(B) 55 lbf
(C) 62 lbf
(D) 80 lbf

49. Measuring industrial performance remains a difficult task. To address this issue, the American Productivity Center has introduced the following integrated measurement criterion for measuring total productivity.

$$\text{total productivity} = (\text{productivity})(\text{price recovery})$$

$$\text{productivity} = \frac{\text{percent change in output quantity}}{\text{percent change in resource quantity}}$$

$$\text{price recovery} = \frac{\text{percent change in output price}}{\text{percent change in resource cost}}$$

An automotive parts supplier generated the data in the following table.

		period n	period $n+1$
output	units produced	1200	1400
	price per unit ($)	54.08	57.23
resource	time worked (hr)	8200	7800
	cost per hour ($)	12.05	13.15

The supplier's overall productivity improvement from period n to period $n+1$ is most nearly

(A) 1.8%
(B) 12%
(C) 19%
(D) 120%

50. A plant manager has been asked by the company board of directors to provide a summary of the plant's productivity gains over the past five years. Though there are many ways in which productivity can be measured, this company uses the variable cost per unit produced. Data on the company's performance over the past five years is given in the following table.

	year 1	year 2	year 3	year 4	year 5
total variable cost	$150,000	$170,000	$180,000	$210,000	$220,000
total units produced	120,000	150,000	155,000	195,000	210,000

The average change in variable cost per unit over the past five years is most nearly

(A) 4%
(B) 8%
(C) 10%
(D) 16%

51. Consider the following work sampling study for a piece of production equipment based upon a random sample of 100 observations.

	number of times observed
machine idle—no product	5
machine operating	85
machine idle—downtime and repair	10

Suppose that an accuracy of 0.05 is required for the study with a confidence of 95%. How many additional observations must be made?

(A) 0
(B) 50
(C) 96
(D) 150

52. Consider the following time study summary for a particular job. Element 2 is a machine-controlled element. Element 4 occurs every third cycle. All times are in minutes.

element	cycles timed	average (min)	standard deviation (min)	occurrences per cycle	rating (%)
1	30	0.246	0.031	1	110
2	30	1.214	0.001	1	100
3	30	0.252	0.022	1	110
4	10	1.682	0.015	$\frac{1}{3}$	90

Most nearly, what is the allowed time, in minutes, for this job before allowances?

(A) 1.8
(B) 2.3
(C) 3.1
(D) 3.2

53. Consider the time study summary for Prob. 52.

An accuracy of ± 0.01 min is required for each element. How many additional cycles must be timed for 95% accuracy?

(A) 0
(B) 7
(C) 10
(D) 50

54. Which of the following contains only those tools typically used for conducting a graphical analysis of worker productivity?

(A) operation process charts, cash flow diagrams, and critical path diagrams
(B) value-added charts, flow diagrams, and force/stress diagrams
(C) cash flow diagrams, PERT diagrams, and flow process charts
(D) operation process charts, flow process charts, and left-hand/right-hand charts

55. 25 samples of size 5 were drawn from a population, and the average range for a measured diameter was determined to be 8.960. Most nearly, what is the estimate of the population standard deviation?

(A) 3.9
(B) 5.0
(C) 6.0
(D) 9.0

56. The permanent control chart limits for an R-chart used to track the sample ranges described in Prob. 55 should be set most nearly at

(A) 18.91, 0
(B) 18.94, 0
(C) 19.81, 0
(D) 19.90, 0

57. The Malcolm-Baldrige Award Program

(A) is a relatively easy program to implement in a manufacturing/service company
(B) is available only to manufacturing companies
(C) evaluates competitors according to a seven-category criteria
(D) is a worldwide competition

58. Many quality management tools provide a systems perspective about the issues that affect product quality. Which of the following tools are NOT typically used in a quality problem analysis?

(A) cash flow diagrams, from-to chart
(B) fish-bone diagrams, histograms
(C) QFD diagram, control charts
(D) scatter diagrams, flow charts

59. A company has produced 10 lots of parts from which it has sampled the weight with a sample size of 6 from each lot. $\overline{X}_i$ and s_i for each sample i is given in the following table. Calculate the upper and lower control limits for the s-chart, and decide whether or not the production process is in control with respect to the standard deviation.

sample, i	$\overline{X}_i$	s_i
1	104	9
2	106	4
3	105	10
4	104	5
5	106	9
6	105	14
7	101	11
8	109	20
9	109	15
10	102	11

(A) $\text{UCL}_s = 35.28$, $\text{LCL}_s = 0.00$, not in control
(B) $\text{UCL}_s = 21.38$, $\text{LCL}_s = 0.32$, in control
(C) $\text{UCL}_s = 18.53$, $\text{LCL}_s = 3.07$, in control
(D) $\text{UCL}_s = 18.53$, $\text{LCL}_s = 3.07$, not in control

60. A customer has specified that the weight of a product should be 10.5 g $\pm$ 2.5 g. Based on 10 samples of 5, $\overline{X}$ and $\overline{s}$ are approximately 10.00 and 1.00. Calculate the process capability index (C_{pk}) for this process and determine if it is capable of producing a product that conforms to the customer's specification.

(A) 0.627, not capable
(B) 0.940, capable
(C) 1.064, not capable
(D) 1.064, capable

SOLUTIONS

1. Use the uniform series compound amount factor to find the future worth of the deposits.

F = future equivalent value of a cash flow
   or series of cash flows

$\qquad = A(F/A, i, n)$

A = uniform series of end of
   compounding period cash flows

$\qquad = \$200$

i = interest rate per compounding period

$\qquad = \dfrac{12\%}{12 \text{ compounding periods per year}}$

$\qquad = 1\%$ per compounding period (month)

n = number of compounding periods

$\qquad = (20 \text{ years})(12 \text{ compounding periods per year})$

$\qquad = 240$ compounding periods

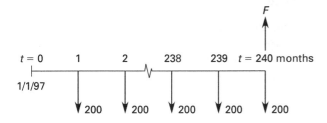

$$F = (\$200)(F/A, 1\%, 240)$$

$$= (\$200)\left(\frac{(1 + 0.01)^{240} - 1}{0.01}\right)$$

$$= \$197{,}851 \quad (\$198{,}000)$$

The answer is D.

2. Use the capital recovery factor to find the monthly car payment.

A = uniform series of end of period cash flows

$\qquad = P(A/P, i, n)$

P = present equivalent value of a cash flow
   or series of cash flows (loan amount)

$\qquad = \$8000$

i = interest rate per compounding period

$\qquad = \dfrac{9\%}{12 \text{ compounding periods per year}}$

$\qquad = 0.75\%$ per compounding period

n = number of compounding periods

$\qquad = (4 \text{ years})(12 \text{ compounding periods per year})$

$\qquad = 48$ compounding periods

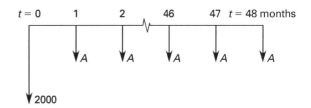

$$A = (\$8000)(A/P, 0.75\%, 48)$$

$$= (\$8000)\left(\frac{(0.0075)(1 + 0.0075)^{48}}{(1 + 0.0075)^{48} - 1}\right)$$

$$= \$199.20 \quad (\$200)$$

The answer is D.

3. Equate the equivalent worth of the two options and solve for the interest rate.

Option 1, receive \$125,000 now ($P_0 = \$125{,}000$).

Option 2, receive \$50,000 now and \$50,000 each year for two years.

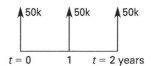

Use the uniform series present worth factor.

$$P_0 = \$50{,}000 + A(P/A, i, n)$$
$$A = \$50{,}000$$
$$n = 2$$
$$P_0 = \$50{,}000 + (\$50{,}000)(P/A, i, 2)$$
$$= \$50{,}000 + (\$50{,}000)\left(\frac{(1 + i)^2 - 1}{i(1 + i)^2}\right)$$

Equate the present worth of the options and solve for the interest rate.

$$\$125{,}000 = \$50{,}000 + (\$50{,}000)\left(\frac{(1 + i)^2 - 1}{i(1 + i)^2}\right)$$

$$1.5 = \left(\frac{(1 + i)^2 - 1}{i(1 + i)^2}\right) = (P/A, i, 2)$$

By trial and error, a range for i can be determined. Then, use linear interpolation to approximate i.

interest rate	$(P/A, i, 2)$
20%	1.5278
i	1.5000
25%	1.4400

$$\frac{25\% - i}{1.44 - 1.5} = \frac{25\% - 20\%}{1.44 - 1.5278}$$

$$i = 0.2158 \quad (22\%)$$

The answer is C.

4.

cost factor (piece)	brass-copper alloy	plastic molding
casting	(25 lbm)($3.35/lbm) = $83.75	(16 lbm)($7.40/lbm) = $118.40
machining	$6.00	0.00
weight penalty	(25 lbm − 16 lbm)($4/lbm) = $36.00	0.00
total cost	$125.75	$118.40

The plastic molding should be selected to save $125.75 − $118.40 = $7.35 over the lifecycle of each radiator.

The answer is C.

5. Determine the cost of producing 15,000 pieces for each tool material.

$$\text{total cost} = (\text{tool cost})(\text{no. of tools needed})$$
$$+ \left(\frac{\text{labor cost}}{\text{hr}}\right)(\text{hr needed})$$

$$\text{tools needed} = \frac{\text{pieces required}}{\left(\dfrac{\text{pieces}}{\text{hr}}\right)\left(\dfrac{\text{tool life in hr}}{\text{tool}}\right)}$$

$$\text{hours needed} = \text{production time}$$
$$+ \text{ tool changing time}$$
$$= \frac{\text{pieces required}}{\dfrac{\text{pieces}}{\text{hr}}}$$
$$+ \left(1 \ \frac{\text{hr}}{\text{tool}}\right)(\text{no. of tools needed})$$

For material A,

$$\text{tools needed} = \frac{15{,}000 \text{ pieces}}{\left(100 \ \dfrac{\text{pieces}}{\text{hr}}\right)\left(50 \ \dfrac{\text{hr}}{\text{tool}}\right)}$$
$$= 3 \text{ tools}$$

$$\text{hours needed} = \frac{15{,}000 \text{ pieces}}{100 \ \dfrac{\text{pieces}}{\text{hr}}} + \left(1 \ \frac{\text{hr}}{\text{tool}}\right)(3 \text{ tools})$$
$$= 153 \text{ hr}$$

$$\text{total cost} = \left(100 \ \frac{\$}{\text{tool}}\right)(3 \text{ tools})$$
$$+ \left(18 \ \frac{\$}{\text{hr}}\right)(153 \text{ hr})$$
$$= \$3054$$

For material B,

$$\text{tools needed} = \frac{15{,}000 \text{ pieces}}{\left(75 \ \dfrac{\text{pieces}}{\text{hr}}\right)\left(25 \ \dfrac{\text{hr}}{\text{tool}}\right)}$$
$$= 8 \text{ tools}$$

$$\text{hours needed} = \frac{15{,}000 \text{ pieces}}{75 \ \dfrac{\text{pieces}}{\text{hr}}} + \left(1 \ \frac{\text{hr}}{\text{tool}}\right)(8 \text{ tools})$$
$$= 208 \text{ hr}$$

$$\text{total cost} = \left(30 \ \frac{\$}{\text{tool}}\right)(8 \text{ tools})$$
$$+ \left(18 \ \frac{\$}{\text{hr}}\right)(208 \text{ hr})$$
$$= \$3984$$

Material A should be selected. The savings are $3984 − $3054 = $930.

The answer is A.

6. To minimize project cost, set the first derivative of the project cost equation (with respect to project length) equal to zero and solve for the project length, T.

$$\frac{d(\text{project cost})}{dT} = -\frac{\$4000 \text{ mo}}{T^2} + \frac{\$500T}{\text{mo}^2} = 0$$
$$\$500T^3 = \$4000 \text{ mo}^3$$
$$T = \sqrt[3]{8 \text{ mo}^3}$$
$$= 2 \text{ mo}$$

To find the expected project cost, substitute $T = 2$ months into the project cost equation.

$$\text{project cost} = \$5000 + \frac{\$4000 \text{ mo}}{T} + \frac{\$250T^2}{\text{mo}^2}$$
$$= \$5000 + \frac{\$4000 \text{ mo}}{2 \text{ mo}}$$
$$+ \frac{(\$250)(2 \text{ mo})^2}{\text{mo}^2}$$
$$= \$8000$$

The answer is C.

7. To calculate the benefit-cost ratio for the new Colorado State Park, first classify the benefits and costs.

$$\text{benefits} = (\$3 \text{ per person entry fee} + \$15 \text{ per person positive economic impact})(500{,}000 \text{ people per year})$$
$$= \$9{,}000{,}000 \text{ per year}$$

$$\text{costs} = \$20{,}000{,}000 \text{ development cost at } t_0 \text{ and } \$2{,}000{,}000 \text{ per year operating expense}$$

Calculate the present value of the benefits (P_{benefits}) and the present value of the cost (P_{costs}).

$$P_{\text{benefits}} = \$9{,}000{,}000 \ (P/A, 6\%, 20)$$
$$= \$9{,}000{,}000 \ (11.4699)$$
$$= \$103{,}229{,}290.97$$

$$P_{\text{costs}} = \$20{,}000{,}000 + 2{,}000{,}000 \ (P/A, 6\%, 20)$$
$$= \$20{,}000{,}000 + 2{,}000{,}000 \ (11.4699)$$
$$= \$42{,}939{,}842.44$$

Finally, take the ratio of the P_{benefits} and P_{costs}.

$$B/C \text{ ratio} = \$103{,}229{,}290.97/\$42{,}939{,}842.44$$
$$= 2.4040 \quad (2.4)$$

The answer is C.

8. The pay-back period (PBP) is based on calculating the break-even point, that is, the time before an initial investment (I) is recovered. The pay-back period when revenues (R) and costs (C) are constant is

$$\text{PBP} = \frac{I}{R - C}$$
$$= \frac{\$80{,}000}{25{,}000 \ \dfrac{\$}{\text{yr}} - 10{,}000 \ \dfrac{\$}{\text{yr}}}$$
$$= 6 \text{ yr}$$

The answer is B.

9. The incremental investment from A_1 to A_2 is $500, and the incremental return is $550. The incremental rate of return is

$$\frac{\$550}{\$500} - 1 = 0.10 \quad (10\%)$$

The answer is A.

10. From the F-distribution table,

$$F^*_{4,16,0.05} = 3.01$$

The mean squares for errors is

$$\frac{116.08}{16} = 7.255$$

The mean squares for replications is

$$\frac{122.46}{4} = 30.62$$
$$F = \frac{30.62}{7.255} = 4.22$$

The mean squares for treatments is

$$\frac{126.92}{4} = 31.73$$
$$F = \frac{31.73}{7.255} = 4.37$$

The treatments are greater than F^* and are significant at 0.05. However, the replications are also significant. The results have a lack-of-fit problem, and a conclusion cannot be made.

The answer is C.

11. From the problem table,

$$F^*_{1,8,0.05} = 5.32$$

The mean squares for errors is

$$\frac{32}{8} = 4.00$$

The mean square for A is

$$\frac{108.5}{1} = 108.50$$
$$F = \frac{108.50}{4.00} = 27.13$$

The mean square for B is

$$\frac{61.22}{1} = 61.22$$
$$F = \frac{61.22}{4.00} = 15.31$$

The mean square for AB is

$$\frac{9.60}{1} = 9.60$$
$$F = \frac{9.60}{4.00} = 2.40$$

F values for factors A and B are greater than F^*. Factors A and B are both significant at a 0.05 level.

The answer is B.

12. For most engineering properties, the high-order cross terms are usually nonexistent, and they can be used as error terms if more experiments are not feasible. One should pool the high-order cross terms as error terms, then conduct the analysis of variance.

The answer is C.

13. Let t = service time in hours. $f(t)$ is exponentially distributed.

$$f(t) = \lambda e^{-\lambda t} \quad [t > 0]$$

$$\lambda = \frac{1}{\mu}$$

$$\mu = \text{population mean}$$

Let the sample mean be an estimate of the population mean.

The estimate for μ is the sample average, 1.25.

$$\lambda = \frac{1}{1.25} = 0.8$$

$$P(t > 2) = \int_2^\infty 0.8 e^{-0.8t} dt = -e^{-0.8t}\Big|_2^\infty$$

$$= 0 + e^{-1.6}$$

$$= 0.2018 \quad (0.20)$$

The answer is D.

14. From the problem statement, H_0 is $\mu \geq 450$, and H_1 is $\mu < 450$.

Accept hypothesis H_0 if

$$\frac{\bar{x} - 450}{\frac{s}{\sqrt{15}}} \geq -t_{0.05,14} = -1.761$$

$$t = \frac{445 - 450}{\frac{10}{\sqrt{15}}} = -1.936$$

$$-1.936 < -1.761$$

Reject H_0.

The answer is B.

15. Use the information given in Sol. 14 to find the p-value.

$$t = \frac{\bar{x} - 450}{\frac{s}{\sqrt{15}}}$$

$$= \frac{445 - 450}{\frac{10}{\sqrt{15}}}$$

$$= -1.936$$

$$p(t_{n=14} < -1.936) = x$$

From the t-distribution table with $15 - 1 = 14$ degrees of freedom,

$$0.05 \longrightarrow -1.761$$
$$x \longrightarrow -1.936$$
$$0.025 \longrightarrow -2.145$$

$$x = 0.05 - \left(\frac{-1.761 - (-1.936)}{-1.761 - (-2.145)} \right)(0.025)$$

$$= 0.0386 \quad (0.040)$$

The answer is D.

16. A 95% confidence interval on mean transporter utilization means there is a 95% chance the population (or true) mean transporter utilization lies within the given interval.

The answer is B.

17. Use the Poisson distribution to approximate the binomial probability.

$$P(\text{accept lot}|P, n, c) = P(x = 0, n = 3, nP = 3P)$$

$$P = \text{population process proportion defective}$$

$$P(x = 0) = \frac{e^{-nP}(nP)^x}{x!} = \frac{e^{-(3)(0.30)}((3)(0.30))^0}{0!}$$

$$= 0.407 \quad (40\%)$$

The answer is C.

18. This problem uses the binominal distribution to calculate the probability of the number of damaged products. Use the equation

$$P_n(x) = \frac{n!}{x!(n-x)!} p^x q^{n-x}$$

$$n = \text{number samples} = 10$$
$$x = \text{number damaged} = 2$$
$$p = \text{probability of damage} = 0.1$$
$$q = 1 - p = 1 - 0.1 = 0.9$$

The probability that two damaged products will be found from a sample of 10 is

$$P_n(x) = \left(\frac{10!}{2!(10-2)!} \right)(0.1)^2(0.9)^{(10-2)}$$

$$= 0.1937 \quad (0.194)$$

The answer is B.

19. The equation for balance delay is

$$\frac{(\text{no. of stations})(\text{cycle time}) - \text{sum of task times}}{(\text{no. of stations})(\text{cycle time})}$$

The spreadsheet formula is

$$(B1 * B2 - SUM(A1 \ldots A5))/B1 * B2$$

The answer is D.

20. The first "While" statement in the pseudocode reads the first seven data points from the data file. The second "While" statement reads the next five data points from the data file and adds them together. It also increments N by 1 each time. This yields the value $F = 1100 + 1150 + 1300 + 1320 + 1440 = 6310$ and $N = 6$. The code then sets $P = F/(N-1)$, which gives $P = 6310/5 = 1262$. The value 1262 satisfies the second "If" statement. Therefore, the value of SF is 8.

The answer is B.

21. This problem can be solved by the Hungarian method for assignment problems by giving the assignments that are not possible a large associated cost, such as \$1000. For this problem, the cost matrix is

$$\begin{bmatrix} 20 & 1000 & 20 & 8 \\ 14 & 16 & 8 & 12 \\ 36 & 12 & 20 & 28 \\ 28 & 8 & 24 & 1000 \end{bmatrix}$$

The reduced cost matrix is found by subtracting the smallest value in each row from each of the other elements in that row and then, for the resultant matrix, subtracting the smallest element in each column from each of the other elements in the column. For the above matrix, subtract 8 from each element of row one, 8 from each element of row two, 12 from each element of row three, and 8 from each element of row four. The resulting matrix is

$$\begin{bmatrix} 12 & 992 & 12 & 0 \\ 6 & 8 & 0 & 4 \\ 24 & 0 & 8 & 16 \\ 20 & 0 & 16 & 992 \end{bmatrix}$$

From the previous matrix, subtract 6 from each element of column one, and 0 from each element of the other three columns, resulting in the following reduced cost matrix.

$$\begin{bmatrix} 6 & 992 & 12 & 0 \\ 0 & 8 & 0 & 4 \\ 18 & 0 & 8 & 16 \\ 14 & 0 & 16 & 992 \end{bmatrix}$$

The minimum number of lines to cover the zeros for the reduced matrix is three.

$$\begin{bmatrix} 6 & 992 & 12 & 0 \\ 0 & 8 & 0 & 4 \\ 18 & 0 & 8 & 16 \\ 14 & 0 & 16 & 992 \end{bmatrix}$$

The new reduced cost matrix is found by subtracting the smallest uncovered number from each uncovered number, and adding that number to each twice covered number. In the previous matrix, 8 is the smallest uncovered number. The new reduced cost matrix is

$$\begin{bmatrix} 6 & 1000 & 12 & 0 \\ 0 & 16 & 0 & 4 \\ 10 & 0 & 0 & 8 \\ 6 & 0 & 8 & 984 \end{bmatrix}$$

The minimum number of lines to cover the zeros for the new reduced cost matrix is four.

$$\begin{bmatrix} 6 & 1000 & 12 & 0 \\ 0 & 16 & 0 & 4 \\ 10 & 0 & 0 & 8 \\ 6 & 0 & 8 & 984 \end{bmatrix}$$

The new reduced cost matrix gives the optimal solution, which is obtained from the zero cells. Operator 1 must be assigned to machine 4, and operator 4 must be assigned to machine 2. After these assignments, operator 3 must be assigned to machine 3, and operator 2 to machine 1. The optimal assignment is then 1-4, 2-1, 3-3, and 4-2. The associated cost for this assignment from the original cost matrix is

$$8 + 14 + 20 + 8 = 50$$

The answer is C.

22. To find the solution given by the northwest corner rule, the transportation matrix must be formed. There is one row for each source and one column for each destination. There is also a row for demand and a column for supply. To use the northwest corner rule, start in the upper left corner and assign to that cell as much as possible to meet either the supply or demand for that row or column. If the supply is met, go to the adjacent cell in the column and repeat the process. If the demand is met, go to the adjacent cell in the row and repeat the process. If both the demand and supply are met, put a 0 in either the adjacent column cell or the adjacent row cell. If a 0 is put in the adjacent column cell, move to the adjacent row cell. If the 0 is put in the adjacent row cell, move to the adjacent column cell. Continue until each row and column has its supply and demand met. This is shown as follows.

source	destinations					supply
	1	2	3	4	5	
1	5	15				20
2		0	40	0		40
3				10		10
4				25	35	60
demand	5	15	40	35	35	

5 is put in cell $(1, 1)$, meeting the demand of column 1. Next, move to the adjacent row cell $(1, 2)$ and assign the value of 15. This meets both the supply and demand of row 1 and column 2, respectively. Since both are met, put a 0 in the adjacent column cell $(2, 2)$, then move to the row cell $(2, 3)$, which is adjacent to $(2, 2)$. For this cell, a value of 40 is assigned, which again meets both the supply and demand of the corresponding row and column (i.e., row 2 and column 3). Since both are met, put a 0 in the adjacent row cell $(2, 4)$, then move to the adjacent column cell $(3, 4)$. A value of 10 is assigned to this cell, meeting the supply of row 3. Then move to the adjacent column cell $(4, 4)$ and assign a value of 25, meeting the demand of column 4. Since the column demand is met, move to the adjacent row cell $(4, 5)$ and assign a value 35, meeting the supply and demand for row 4 and column 5, completing the initial solution since all the demands and supplies are now met.

Using the costs in the corresponding source-destination matrix, the corresponding objective function value is

$$(5)(10) + (15)(15) + (0)(4) + (40)(12) + (0)(23)$$
$$+ (10)(28) + (25)(51) + (35)(29)$$
$$= 3325 \quad (3300)$$

The answer is D.

23. This problem can be viewed as a discrete Markov process. To find the expected value of stock 2, the steady-state probabilities of the values of stock 2 must be known. Let state 1 represent $20 and state 2 represent $35. The corresponding one-step probability transition matrix for stock 2 is

$$\mathbf{P} = \begin{bmatrix} 0.9 & 0.1 \\ 0.2 & 0.8 \end{bmatrix}$$

To find the steady-state probabilities, π_i, solve the system of equations given by $\boldsymbol{\pi}\mathbf{P} = \boldsymbol{\pi}$, which is equivalent to $\boldsymbol{\pi}(\mathbf{P} - \mathbf{I}) = \mathbf{0}$. This system can be written as

$$(\pi_1 \ \pi_2) \begin{pmatrix} -0.1 & 0.1 \\ 0.2 & -0.2 \end{pmatrix} = (0 \ 0)$$

This yields the following system of equations.

$$-0.1\pi_1 + 0.2\pi_2 = 0$$
$$0.1\pi_1 - 0.2\pi_2 = 0$$

This system reduces to $-0.1\pi_1 + 0.2\pi_2 = 0$, which has the solution $\pi_1 = 2\pi_2$. The normalizing equation $(\Sigma\pi_i = 1)$ yields $2\pi_2 + \pi_2 = 1$, which implies $\pi_2 = \frac{1}{3}$ and $\pi_1 = \frac{2}{3}$. The expected value of the stock is then given by

$$\left(\tfrac{2}{3}\right)(\$20) + \left(\tfrac{1}{3}\right)(\$35) = \$25$$

The answer is B.

24. Since the arrivals occur according to a Poisson process, the interarrival times are exponential (M). Even though the service (machining) time is exponential for each machine, service times are not exponential because the service occurs in three consecutive stages. Therefore, service is actually the total of the three stages. However, because the stages are consecutive and each stage has exponential service times, service times are distributed according to an Erlang-3 distribution (E_3). The capacity of the buffer is the limiting factor on the number of products in the system. Therefore, the capacity of the total number of products in the system is 11:1 being machined and 10 waiting.

The answer is D.

25. Interarrival times and service times are exponentially distributed. There is also room for a large number of people in the food court, and people can wait outside the court, implying that the capacity of the system is infinite (very large). Therefore, this is an M/M/2 model. For this model, the number of people waiting in line is given by

$$L_q = \frac{2\rho^3}{1 - \rho^2}$$

From the problem statement, $\lambda = 30/\text{hr}$ is the arrival rate and $\mu = 20/\text{hr}$ is the service rate for each teller.

The server utilization is

$$\rho = \frac{\lambda}{s\mu} = \frac{30\,\frac{1}{\text{hr}}}{(2)\left(20\,\frac{1}{\text{hr}}\right)} = 0.75$$

The number of people waiting in line is

$$L_q = \frac{(2)(0.75)^3}{1 - (0.75)^2} = 1.93 \text{ people} \quad (2 \text{ people})$$

The answer is C.

26. Management by objectives best represents the company's performance appraisal system. This system sets up specific objectives for each individual in the organization with respect to overall organizational objectives, and it holds each individual accountable for achieving these objectives.

Value based management focuses on aligning an organization around how to create value (typically shareholder value). Performance management is used to determine if a company is achieving its overall performance objectives. Change management focuses on the need for organizational change, not individual employee performance.

The answer is D.

27. Gain share is calculated by determining the number of additional units produced over the standard, and then adding the base pay and the additional 50-50 share to obtain the total pay. To find how much the employee will make, first find the base number of units produced at the standard.

$$\text{standard units produced} = \frac{60 \text{ min}}{2.5 \dfrac{\text{min}}{\text{unit}}}$$
$$= 24 \text{ units}$$

The problem states that an employee averages 30 units per day, so subtract the 24 standard units from the 30 actual units to find that 6 units are produced over the standard.

A 50-50 gain share implies that the worker receives an additional 50% for each unit produced over the standard. The standard base pay is $12 per hour, or $0.20 per minute, so the standard pay per unit is

$$\left(2.5 \frac{\text{min}}{\text{unit}}\right)\left(0.20 \frac{\$}{\text{min}}\right) = \$0.50/\text{unit}$$

The total pay can now be calculated as follows.

$$\text{total pay} = (\text{hourly rate} + 50\% \text{ gain share})$$
$$\times (\text{total hours worked})$$
$$= \left(12.00 \frac{\$}{\text{hr}} + (0.5)\left(6 \frac{\text{unit}}{\text{hr}}\right)\left(0.50 \frac{\$}{\text{unit}}\right)\right)$$
$$\times \left(8 \frac{\text{hr}}{\text{day}}\right)$$
$$= \$108.00/\text{day}$$

The answer is D.

28. The critical path is defined as the longest path through the network. Calculate the path duration for each possible solution.

Activity sequence A is

$$1\text{-}3, 3\text{-}6 = 3 + 16 = 19$$

Activity sequence B is

$$1\text{-}2, 2\text{-}5, 5\text{-}6 = 4 + 10 + 2 = 16$$

Activity sequence C is

$$1\text{-}2, 2\text{-}4, 4\text{-}6 = 4 + 6 + 8 = 18$$

Activity sequence D is

$$1\text{-}3, 3\text{-}4, 4\text{-}6 = 3 + 2 + 8 = 13$$

The answer is A.

29. The early start schedule is defined as the schedule that has each activity starting as soon as possible, or in other words, has the minimum path time to an activity.

The minimum path time calculations are shown in the illustration and result in the following early start schedule.

$A_{1\text{-}2} = 0$; $A_{1\text{-}3} = 0$; $A_{2\text{-}4} = 4$; $A_{2\text{-}5} = 4$; $A_{3\text{-}4} = 3$; $A_{3\text{-}6} = 3$; $A_{4\text{-}6} = 10$; $A_{5\text{-}6} = 14$

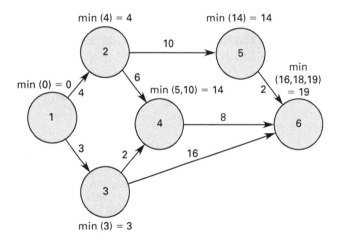

The answer is A.

30. Theory Z is the motivation theory most closely described by the president's new approach. It assumes that workers are motivated, that management's role is to involve workers in decision making, and that the company will provide lifelong opportunities for service.

Theory W doesn't exist. Theory X assumes workers are at best passive regarding a company's objectives and, therefore, management needs to reward and control their activities. Theory Y assumes workers are inherently motivated so management only needs to set up desirable work conditions.

The answer is D.

31. Calculate the probability of each path being completed in 21 days.

$$P\{T \le X\} = P\{z \le (X - \mu)/\sigma\}$$

Determine the value for z using a standard normal table.

$$P\{T \le 21\} = P\{z \le (21 - \mu)/\sigma\}$$
$$\text{probability of critical path} = P\{z \le (21 - 20)/2\}$$
$$= P(z \le 0.5)$$
$$= 0.691$$

$$\text{probability of}$$
$$\text{non-critical path} = P\{z \le (21 - 18.5)/6\}$$
$$= P(z \le 0.4167)$$
$$= 0.664$$

The probability of completing the project is the maximum probability for the two paths (0.691, 0.664), or 0.691.

The answer is D.

32. Use the following equations to determine the number of machines required.

$$F = \text{total number of machines ceiling} = \sum F_k$$
$$= F_1 + F_2$$
$$F_k = \frac{S_{i,k} Q_i}{EHR \prod_{i}(1 - P_{i,k})}$$
$$= \text{number of machines for operation } k$$

$S_i = $ standard time required per part i for operation k

$Q_i = $ production quantity per time period for part i

$E = $ actual performance of machine

$H = $ available time units per time period

$R = $ machine k availability

$P_{i,k} = $ scrap rate for machine k part i

$$F_1 = \frac{(2)(5000)}{(0.9)\big((5)(10)(60)\big)(0.9)(1 - 0.03)(1 - 0.05)}$$
$$= 4.47$$
$$F_2 = \frac{(4)(5000)}{(0.9)\big((5)(10)(60)\big)(0.95)(1 - 0.05)}$$
$$= 8.21$$
$$F = \text{ceiling }(4.47 + 8.21) = \text{ceiling }(12.68) = 13$$

The answer is C.

33. The machining time required for a single pass is

$t = $ time

$f = $ feed

$s = $ spindle speed

$L = $ length of cut

$$t = \frac{L}{fs} = \frac{24 \text{ in}}{\left(0.012 \, \dfrac{\text{in}}{\text{rev}}\right)\left(400 \, \dfrac{\text{rev}}{\text{min}}\right)}$$
$$= 5.0 \text{ min}$$

The answer is B.

34. The time to cut twelve 4 in pieces is

$t = $ time

$n = $ number of pieces

$p = $ set-up time

$A = $ cut area

$r = $ cutting rate

$$t = n\left(p + \frac{A}{r}\right)$$
$$= (12)\left((15 \text{ sec})\left(\frac{1 \text{ min}}{60 \text{ sec}}\right) + \frac{\pi \left(\dfrac{3 \text{ in}}{2}\right)^2}{12 \, \dfrac{\text{in}^2}{\text{min}}}\right)$$
$$= 10.07 \text{ min} \quad (10 \text{ min})$$

The answer is C.

35. Reducing the number of components, including the use of snap fasteners, is the most appropriate design for assembly/manufacturing guidelines for improving the product as described.

Rule 1 does not apply because there are no flexible components (springs, wires, etc.), rule 2 does not apply because the current operations are already as simple as possible, and rule 4 does not apply because there is no problem with confusing insertion geometry in this product.

The answer is B.

36. The makespan is determined by the time at which the last job in the sequence finishes at the second station. Starting with the first job in the sequence, determine the start and finish time of the painting operation and the start and finish time of the inspection operation. The next job in the sequence will begin its painting operation immediately following the finish of the previous job in the sequence. A job will begin the inspection operation when its painting operation is complete and the inspection station has finished the preceding job in the sequence. The following table shows the scheduled start and finish times of all jobs at each of the stations.

	painting operation		inspection operation	
	start time	finish time	start time	finish time
job 3	0	15	15	23
job 2	15	23	23	28
job 1	23	35	35	44
job 4	35	42	44	48

Job 4, the last job in the sequence, is scheduled to complete its inspection operation at time 48. Thus, the makespan of the sequence $\{3, 2, 1, 4\}$ is 48.

The answer is D.

37. The general formula for determining the economic order quantity is

$$Q^* = \sqrt{\frac{2K\lambda}{h}}$$

K = cost to place an order = \$15

h = holding cost per unit per period

   = \$0.10/unit-wk

λ = average demand per period

   = 750 units/wk

$$Q^* = \sqrt{\frac{(2)(\$15)\left(750 \, \dfrac{\text{units}}{\text{wk}}\right)}{0.10 \, \dfrac{\$}{\text{unit-wk}}}}$$

   = 474.3 units   (470 units)

The answer is C.

38. The general equation for determining the minimum workforce requirement is

W = minimum number of workers per shift

$$= \frac{D_w T_p}{S_w H}$$

D_w = average weekly demand = 10,000 units/wk

T_p = production time per unit = 0.25 hr/unit

S_w = number of shifts per week

$$= \left(2 \, \frac{\text{shifts}}{\text{day}}\right)\left(5 \, \frac{\text{days}}{\text{wk}}\right)$$

$$= 10 \text{ shifts/wk}$$

H = number of productive hours per person

$$= 8.5 \, \frac{\text{hr}}{\text{person}} - 1.5 \, \frac{\text{hr}}{\text{person}}$$

$$= 7 \text{ hr/person}$$

$$W = \frac{\left(10{,}000 \, \dfrac{\text{units}}{\text{wk}}\right)\left(0.25 \, \dfrac{\text{hr}}{\text{unit}}\right)}{\left(10 \, \dfrac{\text{shifts}}{\text{wk}}\right)\left(7 \, \dfrac{\text{hr}}{\text{person}}\right)}$$

$$= 35.7 \text{ people/shift} \quad (36 \text{ people/shift})$$

A minimum workforce of 36 people per shift is needed to meet the demand without overtime. If only 35 people per shift are used, weekly production will be less than weekly demand.

The answer is D.

39. The equation to calculate the number of kanban cards is

$$n = \frac{DL(1 + \alpha)}{C}$$

D = demand per unit = 500 units

L = average lead time for a kanban quantity

   (fraction of a day) = 2.0

C = container capacity = 100 units

α = safety factor = 10%, or 0.10

$$n = \frac{(500 \text{ units})(2)(1 + 0.10)}{100 \text{ units}}$$

$$= 11$$

The answer is D.

40. To calculate the total material handling cost impact, $\Delta TC_{2,4}$, it is necessary to calculate the change in cost resulting from exchanging the two cells (C2 and C4). Let w be the weighting factor determined as the amount of flow between cells, and let d be the distance between cells.

$$\Delta \text{TC}_{2,4} = (w_{1,2} - w_{1,4})(d_{1,2} - d_{1,4})$$
$$+ (w_{2,2} - w_{2,4})(d_{2,2} - d_{2,4})$$
$$+ (w_{3,2} - w_{3,4})(d_{3,2} - d_{3,4})$$
$$+ (w_{4,2} - w_{4,4})(d_{4,2} - d_{4,4})$$
$$+ (w_{5,2} - w_{5,4})(d_{5,2} - d_{5,4})$$
$$+ (w_{6,2} - w_{6,4})(d_{6,2} - d_{6,4})$$
$$- 2w_{2,4}(d_{2,4})$$
$$= (150 - 50)(30 - 10) + (0 - 80)(0 - 20)$$
$$+ (50 - 30)(20 - 20) + (0 - 0)(10 - 30)$$
$$+ (10 - 100)(10 - 10) + (80 - 0)(20 - 0)$$
$$- (2)(80)(20)$$
$$= 2000$$

The answer is C.

41. Products are assigned locations based on their throughput (T) to number of location (S) ratio (T/S). The order (or priority) is according to a non-increasing T/S ratio.

product	no. of locations required S	total loads moved per day T	T/S
1	10	200	$\frac{200}{10} = 20$
2	16	160	$\frac{160}{16} = 10$
3	8	240	$\frac{240}{8} = 30$
4	6	150	$\frac{150}{6} = 25$

Non-increasing order of T/S is $3, 4, 1, 2$.

The answer is D.

42. Since parts are palletized 5 per pallet, the arrival rate of pallets to the conveyor is 100/5, or 20 pallets per minute. Since the conveyor is only 1.2 m wide, the pallet must be placed on the conveyor with the 1.3 m side parallel with the conveyor. This implies that a minimum 1.3 m footprint is required for each pallet on the conveyor; therefore, the conveyor must go fast enough to clear a space for each new pallet. The minimum speed the conveyor must go is

$$\left(20 \ \frac{\text{loads}}{\text{min}}\right)\left(1.3 \ \frac{\text{m}}{\text{load}}\right) = 26 \text{ m/min}$$

The answer is C.

43. The average utilization for the four forklifts is

$$\rho = \frac{L}{s\mu}$$

ρ = utilization
L = mean arrival rate = 40 moves/hr
s = no. of servers = 4 forklifts
μ = mean service time

$$= \frac{60 \ \frac{\text{min}}{\text{hr}}}{1 \text{ min load} + 1 \text{ min unload} + \frac{240 \text{ m}}{80 \ \frac{\text{m}}{\text{min}}}}$$
$$= 12 \text{ hr}$$
$$\rho = \frac{40 \ \frac{\text{moves}}{\text{hr}}}{(4 \text{ forklifts})(12 \text{ hr})} = 0.8333 \quad (85\%)$$

The answer is C.

44. The product 1 distance is

$$50 \text{ m} + 50 \text{ m} = 100 \text{ m}$$

The product 2 distance is

$$100 \text{ m} + 50 \text{ m} + 50 \text{ m} + 100 \text{ m} = 300 \text{ m}$$

(total flow)(distance)
$$= (100 \text{ loads})(100 \text{ m}) + (200 \text{ loads})(300 \text{ m})$$
$$= 70{,}000 \text{ load-m}$$

The answer is D.

45. To calculate the flow times distance cost ($F \times D$) for each location, the distance between each rectilinear centroid is determined by inspection. Storage A is directly below department 1 so the distance between them is 10 m. Storage A is 10 m below and 10 m to the left of department 2 so the distance between them is 20 m, and so forth.

	dept. 1	dept. 2	dept. 3	dept. 4
storage A	10 m	20 m	10 m	20 m
storage B	20 m	10 m	20 m	10 m

Multiply each distance by the flows between each department to determine which storage location has the lowest $F \times D$ cost.

$F \times D_{\text{storage A}}$

$$= \left(1 \, \frac{\$}{m}\right) \begin{pmatrix} \left(250 \, \dfrac{\text{units}}{\text{yr}}\right)\left(10 \, \dfrac{m}{\text{unit}}\right) \\ + \left(400 \, \dfrac{\text{units}}{\text{yr}}\right)\left(20 \, \dfrac{m}{\text{unit}}\right) \\ + \left(300 \, \dfrac{\text{units}}{\text{yr}}\right)\left(10 \, \dfrac{m}{\text{unit}}\right) \\ + \left(600 \, \dfrac{\text{units}}{\text{yr}}\right)\left(20 \, \dfrac{m}{\text{unit}}\right) \end{pmatrix}$$

$= \$25{,}500/\text{yr}$

$F \times D_{\text{storage B}}$

$$= \left(1 \, \frac{\$}{m}\right) \begin{pmatrix} \left(250 \, \dfrac{\text{units}}{\text{yr}}\right)\left(20 \, \dfrac{m}{\text{unit}}\right) \\ + \left(400 \, \dfrac{\text{units}}{\text{yr}}\right)\left(10 \, \dfrac{m}{\text{unit}}\right) \\ + \left(300 \, \dfrac{\text{units}}{\text{yr}}\right)\left(20 \, \dfrac{m}{\text{unit}}\right) \\ + \left(600 \, \dfrac{\text{units}}{\text{yr}}\right)\left(10 \, \dfrac{m}{\text{unit}}\right) \end{pmatrix}$$

$= \$21{,}000/\text{yr}$

Storage B has the lowest $F \times D$ cost.

The answer is B.

46. To determine the product classes, the products must first be ranked by the throughput/space (T/S) ratio to determine which products contribute the most warehouse activity. Then, classes are formed by assigning products that contribute the top 50% of activity to class A, 30% of activity to class B, and 20% of activity to class C. (See *Table for Solution 46*.)

From the table, class A is composed of products (4, 11), class B of products (8, 2, 7), and class C of products (5, 1, 10, 9, 12, 3, 6).

The answer is A.

47. The hip breadth while sitting for the 95th percentile from the NCEES Handbook's ergonomics table is 43.7 cm for women, and 40.6 cm for men.

Design for women.

$$43.7 \text{ cm} + 1.2 \text{ cm} = 44.9 \text{ cm} \quad (45 \text{ cm})$$

The answer is D.

48. The NIOSH formula for action limit is

$$\text{AL} = (90)\left(\frac{6}{H}\right)\left(1 - (0.01)(|V - 30|)\right)$$
$$\times \left(0.7 + \frac{3}{D}\right)\left(1 - \frac{F}{F_{\text{max}}}\right)$$

H = horizontal distance of the hand from the body's center of gravity at the beginning of the lift

$\quad = 16 \text{ in}$

Table for Solution 46

product	throughput T	space S	T/S	% total activity	rank	cumulative % activity	product class
4	26.0	3	8.7	$8.7/29.8 = 0.29$	1	0.29	A
11	56.0	9	6.2	$6.2/29.8 = 0.21$	2	0.50	A
8	27.0	6	4.5	$4.5/29.8 = 0.15$	3	0.65	B
2	10.0	4	2.5	$2.5/29.8 = 0.08$	4	0.73	B
7	8.4	5	1.7	$1.7/29.8 = 0.06$	5	0.79	B
5	6.4	4	1.6	$1.6/29.8 = 0.05$	6	0.84	C
1	3.6	3	1.2	$1.2/29.8 = 0.04$	7	0.88	C
10	5.5	5	1.1	$1.1/29.8 = 0.04$	8	0.92	C
9	7.2	8	0.9	$0.9/29.8 = 0.03$	9	0.95	C
12	7.0	10	0.7	$0.7/29.8 = 0.02$	10	0.97	C
3	5.0	12	0.4	$0.4/29.8 = 0.01$	11	0.98	C
6	1.2	4	0.3	$0.3/29.8 = 0.01$	12	0.99	C
totals	163.3	73	29.8				

V = vertical distance from the hands to the floor at the beginning of the lift

$\quad$ = 12 in

D = distance the object is lifted vertically

$\quad$ = 48 in

F = average number of lifts per minute

$\quad$ = 0.4 lifts/min

F_{max} = maximum frequency of lifting that can be sustained over an 8 hr shift

$\quad$ = 12 lifts/min

MPL = maximum permissible limit (in lbf)

$\quad$ = (3)(action limit)

$|V - 30|$ = absolute value of $V - 30$

$$\text{AL} = (90)\left(\frac{6}{16 \text{ in}}\right)(1 - (0.01)(|12 \text{ in} - 30|))$$

$$\times \left(0.7 + \frac{3}{48 \text{ in}}\right)\left(1 - \frac{0.4 \frac{\text{lifts}}{\text{min}}}{12 \frac{\text{lifts}}{\text{min}}}\right)$$

$$= 20.4$$

$$\text{MPL} = (3)(20.4) = 61.2 \quad (62 \text{ lbf})$$

The answer is C.

49. The total productivity is

total productivity

$= $ (productivity)(price recovery)

$= \left(\dfrac{\text{percent change in output quantity}}{\text{percent change in resource quantity}}\right)$

$\quad \times \left(\dfrac{\text{percent change in output price}}{\text{percent change in resource cost}}\right)$

$= \left(\dfrac{\frac{1400 \text{ units}}{1200 \text{ units}}}{\frac{7800 \text{ hr}}{8200 \text{ hr}}}\right)\left(\dfrac{\frac{\$57.23}{\$54.08}}{\frac{\$13.15}{\$12.05}}\right)$

$= 1.189$

There is an 18.9% (19%) total productivity improvement.

The answer is C.

50. Create a table to find the average change in variable cost per unit (VCU) over the past 5 years.

	year 1	year 2	year 3	year 4	year 5
total variable cost	$150,000	$170,000	$180,000	$210,000	$220,000
total units produced	120,000	150,000	155,000	195,000	210,000
$\frac{\text{variable cost}}{\text{unit}}$	1.25	1.133	1.161	1.076	1.047
percentage change from previous year		−9.36%	2.47%	−7.32%	−2.70%

$$\frac{\text{variable cost}}{\text{unit}} = \frac{\text{total variable cost}}{\text{total units produced}}$$

percentage change for year n

$$= \frac{\text{VCU}(n) - \text{VCU}(n-1)}{\text{VCU}(n-1)}$$

The average change in VCU over the five years is

$$\frac{\begin{array}{c}-9.36\% + 2.47\% \\ + (-7.32\%) + (-2.70\%)\end{array}}{4} = -4.23\% \quad (-4\%)$$

The answer is A.

51. The general formula for finding the sample size required for an activity with a stated accuracy and a confidence of 95% is

n = required sample size for the activity

$\quad = \dfrac{(1.96)^2 p(1-p)}{\epsilon^2}$

p = proportion activity observed in the pilot study

1.96 = standardized normal variant for 95% confidence (two-tail analysis since the number may be too large or too small)

ϵ = desired accuracy

For machine idle—no product,

$$p = \frac{5}{100} = 0.05$$

$$n = \frac{(1.96)^2(0.05)(0.95)}{(0.05)^2} = 72.99 \quad (73)$$

For machine operating,

$$p = \frac{85}{100} = 0.85$$

$$n = \frac{(1.96)^2(0.85)(0.15)}{(0.05)^2} = 195.9 \quad (196)$$

For machine idle—downtime and repair,

$$p = \frac{10}{100} = 0.10$$

$$n = \frac{(1.96)^2(0.10)(0.90)}{(0.05)^2} = 138.3 \quad (139)$$

The largest number of required samples is 196.

196 observations required − 100 current observations
   = 96 additional observations

The answer is C.

52. The general formula for allowed time (AT) before allowances is

$$AT = (OT)\left(\frac{R}{100}\right)K$$

OT = observed time
R = objective rating as a percent
K = occurrences per cycle

For element 1,

$$AT = (0.246 \text{ min})\left(\frac{110\%}{100}\right)(1) = 0.2706$$

For element 2,

$$AT = (1.214 \text{ min})\left(\frac{100\%}{100}\right)(1) = 1.2140$$

For element 3,

$$AT = (0.252 \text{ min})\left(\frac{110\%}{100}\right)(1) = 0.2772$$

For element 4,

$$AT = (1.682 \text{ min})\left(\frac{90\%}{100}\right)\left(\frac{1}{3}\right) = 0.5046$$

The total for the job is

$$0.2706 \text{ min} + 1.2140 \text{ min}$$
$$+ 0.2772 \text{ min} + 0.5046 \text{ min} = 2.2664 \text{ min}$$

The answer is B.

53. The general formula for determining the number of cycles to time per element for any element that is not machine-controlled is

$$n = \left(\frac{1.96s}{\epsilon}\right)^2$$

1.96 = normal standardized variant for 95% accuracy (two-tail analysis since the number may be too large or too small)
s = element standard deviation
ϵ = accuracy = ± 0.01

For element 1,

$$n = \frac{(1.96)^2(0.031)^2}{(0.01)^2} = 36.9 \quad (37)$$

Element 2 is machine controlled.

For element 3,

$$n = \frac{(1.96)^2(0.022)^2}{(0.01)^2} = 18.6 \quad (19)$$

For element 4,

$$n = \frac{(1.96)^2(0.015)^2}{(0.01)^2} = 8.64 \quad (9)$$

There are sufficient samples for elements 2, 3 and 4, but element 1 requires seven additional observations. The job must be timed for an additional seven cycles.

The answer is B.

54. Tools for the graphical analysis of worker productivity include operation process charts to analyze if operations are needed, flow process charts to conduct a more detailed analysis of either a product's operations or a worker's processes, and left-hand/right-hand charts to analyze all of the activities that a worker performs.

The answer is D.

55. Use the following formula to find the estimate of the population standard deviation.

$$\hat{\sigma}_X \approx \frac{\overline{R}}{d_2}$$

$\overline{R}$ average of sample ranges = $\Sigma R/K$
K = number of sample ranges = 25
d_2 = constant factor for sample size $n = 5$

d_2 relates the average ranges to an estimate of the population standard deviation $\hat{\sigma}_X$.

$$\hat{\sigma}_X = \frac{8.960}{2.326} = 3.852 \quad (3.850)$$

The answer is A.

56. D_4 is a constant factor to approximate the upper range control limit (UCL) three standard deviations above the R-chart central line.

D_3 is a constant factor to approximate the lower range control limit (LCL) three standard deviations below the R-chart central line.

$$\text{UCL}_R = D_4\overline{R} = (2.114)(8.960) = 18.941$$
$$\text{LCL}_R = D_3\overline{R} = (0)(8.960) = 0$$

The answer is B.

57. The Malcolm-Baldridge Award Program evaluates competitors according to seven categories: leadership; strategic planning; customer and market focus; measurement, analysis, and knowledge management; human resource focus; project management; and business results.

The award is given by the President of the United States, and it is the most prestigious award for quality in the U.S.

The answer is C.

58. Dr. Kaoru Ishikawa defines the seven tools of quality as check sheets, pareto charts, cause and effect diagrams (or fish-bone diagrams), flow charts, histograms, scatter diagrams, and control charts. QFD (quality function deployment) diagrams are another quality tool used to translate customer requirements into product characteristics. Cash flow diagrams are used for economic analysis, and from-to charts are used in material flow analysis, not quality problem analysis.

The answer is A.

59. The equations for the upper control limit (UCL_s) and the lower control limit (LCL_s) are

$$\text{UCL}_s = B_4\overline{s}$$
$$\text{LCL}_s = B_3\overline{s}$$

$$\overline{s} = \frac{\sum_{i=1}^{10} s_i}{n} = \frac{\sum_{i=1}^{10} s_i}{10}$$
$$= \frac{108}{10}$$
$$= 10.8$$

Use $(\overline{X},\ s)$ standard deviation charts for a sample size of 6 to obtain $B_3 = 0.030; B_4 = 1.970$.

$$\text{UCL}_s = (1.97)(10.8) = 21.38$$
$$\text{LCL}_s = (0.03)(10.8) = 0.32$$

Since all values of s_i are within these limits, the process is in control.

The answer is B.

60. The following equation is used to calculate process capability index (C_{pk}), where μ is the process mean, σ is the standard deviation, LSL is the lower specification limit, and USL is the upper specification limit.

$$C_{pk} = \min\left(\frac{\mu - \text{LSL}}{3\sigma},\ \frac{\text{USL} - \mu}{3\sigma}\right)$$
$$\text{LSL} = 10.5\text{ g} - 2.5\text{ g} = 8.0\text{ g}$$
$$\text{USL} = 10.5\text{ g} + 2.5\text{ g} = 13.0\text{ g}$$

Determine σ from c_4 from the statistical quality control tables when $n = 5$.

$$\sigma = \overline{s}/c_4$$
$$= 1.00/0.94$$
$$= 1.064$$
$$C_{pk} = \min\left(\frac{10.00\text{ g} - 8.00\text{ g}}{(3)(1.064)},\ \frac{13.00\text{ g} - 10.00\text{ g}}{(3)(1.064)}\right)$$
$$= \min(0.627\text{ g},\ 0.940\text{ g})$$
$$= 0.627\text{ g}$$

Since, C_{pk} is less than 1.0, the product does not conform to the customer's specifications.

The answer is A.

Practice Exam 2

PROBLEMS

1. A toll bridge across the Missouri River is being considered as a replacement for an existing bridge. Initial construction costs for the structure are estimated to be $22,500,000. Annual operating and maintenance costs are anticipated to be $475,000. Revenues generated from the toll are anticipated to be $3,500,000 in the bridge's first year of operation, with a projected rate of increase of $50,000 per year due to the anticipated annual increase in traffic over the bridge. Assuming zero value for the bridge at the end of its 30 year life and an interest rate of 12% per year, a benefit-cost (B/C) analysis would result in which of the following decisions?

(A) do not build the bridge, $B/C \geq 1$
(B) build the bridge, $B/C \geq 1$
(C) do not build the bridge, $B/C < 1$
(D) build the bridge, $B/C < 1$

2. An automatic insertion machine was purchased for $250,000. It has a modified accelerated cost-recovery system (ACRS) recovery period of 5 years. Using the modified ACRS rates, the depreciation (D) and book value (BV) at the end of year 2 are most nearly

(A) $D_2 = \$50,000$, BV $= \$120,000$
(B) $D_2 = \$80,000$, BV $= \$120,000$
(C) $D_2 = \$83,300$, BV $= \$130,000$
(D) $D_2 = \$111,000$, BV $= \$130,000$

3. Cost estimates for a proposed public facility are being evaluated. Initial construction cost is anticipated to be $120,000, and annual maintenance expenses are expected to be $6500 for the first 20 years and $2000 for every year thereafter. The facility is to be used and maintained for an indefinite period of time. Using an interest rate of 10% per year, the capitalized cost of this facility is most nearly

(A) $160,000
(B) $190,000
(C) $200,000
(D) $270,000

4. Which of the following cost items would NOT be considered an overhead cost?

(A) administrative salaries
(B) property taxes
(C) raw materials
(D) insurance

5. One method of overhead allocation is to assume that overhead is incurred in direct proportion to machine hours used. Suppose that for a future period (say, a quarter) the total overhead cost is expected to be $200,000 and the total machine hours used is expected to be 18,000 hours. For a given unit of production, the machine hours used is expected to be 0.75 hour. The overhead cost for this unit of production is most nearly

(A) $8.33/unit
(B) $11.10/unit
(C) $14.80/unit
(D) $15.00/unit

6. The U-Power-It company manufactures a variety of human-powered vehicles. Its most popular toddler product is a traditional red tricycle. The following cost data apply to the manufacture of the tricycle.

$$\text{fixed cost} = \$8000/\text{yr}$$
$$\text{variable cost} = \$20/\text{tricycle}$$

If the tricycle sells for $40, the number of tricycles that must be sold to make a profit of $10,000 per year is most nearly

(A) 100 tricycles/yr
(B) 500 tricycles/yr
(C) 700 tricycles/yr
(D) 900 tricycles/yr

7. Missouri Power and Light is considering building a new 1000 MW power plant. Previously, the utility had built a 400 MW plant for $350 million when the construction cost index was 121. Given that the current construction cost index is 139, and assuming a power law sizing exponent of 0.75, a reasonable estimate for the cost of the new facility is most nearly

(A) $400 million
(B) $700 million
(C) $800 million
(D) $1000 million

8. A Texas baseball team purchased a $140,000 pitching machine that has a useful and depreciation life of 7 years. If the machine has a salvage value of $20,000 at the end of its life, and straight-line depreciation is used, the book value at the end of year 4 is most nearly

(A) $20,000
(B) $50,000
(C) $60,000
(D) $70,000

9. Last year, two partners doing business as Tiger Ice Cream earned revenues of $800,000. Their total materials and labor expenses were $300,000. Their building and equipment allowed them to take a $100,000 depreciation charge, but they still had to pay their $80,000 loan payment, of which $40,000 was interest. If their effective tax rate is 40%, then the after-tax cash flow for last year is most nearly

(A) $240,000
(B) $280,000
(C) $340,000
(D) $480,000

10. A manufacturer of concrete construction supports is interested in determining the effects of different sand mixtures on the strength, x, of the supports. Four supports are made for each of the four different percentages of mix of sand and then tested for compression strength. The results are shown as follows.

trial j	% sand	sample i 1	2	3	4	$\sum x_i$	$\overline{x}_i$
		compression strength (10,000 psi)					
1	15	7.34	8.41	8.28	7.35	31.38	7.845
2	20	6.8	7.63	7.24	7.46	29.13	7.2825
3	25	7.92	9.43	9.4	9.91	36.66	9.165
4	30	7.87	7.51	7.76	6.4	29.54	7.385
					overall	126.71	31.6775

Conduct an analysis of variance (ANOVA) with $\alpha = 0.05$ to determine if there is an effect due to changing the sand mix. The four trials are independent. The result is most nearly

(A) $F \approx 0.95$, no effect
(B) $F \approx 1.96$, no effect
(C) $F \approx 2.97$, significant effect
(D) $F \approx 7.21$, significant effect

11. The following data from the ASTM Standard A747, CB7Cu-1 present properties of commonly used corrosion-resistant cast steels. Calculate the simple linear regression line, y (hardness in BHN) $= a + bx$, where x is hardened temperatures in degrees Celsius.

x hardened temperature (°C)	y hardness (BHN)
480	375
495	375
550	311
580	277
590	269
605	269

total	3300	1876
mean	550	312.67
sum of squares	1,828,450	599,422

sum of $xy = 1,018,790$

(A) $y = 655.83 - 0.427x$
(B) $y = 844.52 - 0.967x$
(C) $y = 892.55 - 1.05x$
(D) $y = 937.12 - 1.22x$

12. The following data are collected from experiments that were run in an open field on a summer day. An additive was added incrementally to a process every hour starting from 8 A.M. Averages of certain responses were measured shortly after additives were added.

mass of additives (g)	mean responses
10	1.2
15	1.4
20	1.5
25	1.8
30	2.2
35	2.4
40	2.7
45	2.7
50	2.6

What is the best way to analyze these data?

(A) Use ANOVA to conclude that the additive does not have a significant effect on this process.
(B) Use ANOVA to conclude that the additive has a significant effect on this process.
(C) Use linear regression to show the trend of responses due to the additives.
(D) Do nothing, since these data cannot be analyzed due to errors in the experimental process.

13. An engineer has built a simulation model of a small factory, run an experiment with the model using 15 replicates, and determined at 90% confidence that the mean time an entity spends in the system is in the range of 24 min to 28 min. Based on this information, what can the engineer say about entity time-in-system?

(A) The population mean time-in-system must be in the interval 24 min to 28 min.

(B) The largest mean time-in-system for any one of the 15 replicates was 28 min.

(C) If the engineer were to run a 16th replicate, the mean time-in-system for an additional replicate would be in the interval 24 min to 28 min.

(D) The probability that the population mean time-in-system is greater than 28 min can be estimated as 5%.

14. The life in hours of a 60 W bulb is known to be approximately normally distributed with standard deviation $\sigma = 25$ hr. To be 95% confident that the error in estimating the mean life is less than 5 hr, what sample size should be used?

(A) 5
(B) 25
(C) 95
(D) 97

15. An airline knows that 5% of the people making reservations on a certain flight will not show up. Consequently, its policy is to sell 52 tickets for a flight that can only hold 50 passengers. What is the probability that there will be a seat available for every passenger who shows up for the fully reserved flight?

(A) 0.50
(B) 0.74
(C) 0.88
(D) 0.95

16. ACME Manufacturing is testing the diameter consistency of their new baseball bat design. Out of a production batch of 1000, it measures the diameter of 13 bats.

3.12, 3.17, 3.05, 3.19, 3.18, 3.11, 3.16, 3.20, 3.15, 3.16, 3.12, 3.19, 3.15

Calculate the median and sample standard deviation.

(A) 3.15, 0.0398
(B) 3.15, 0.0414
(C) 3.16, 0.0398
(D) 3.16, 0.0414

17. To provide adequate fire protection service, a city has three pumper trucks. Truck 1 is an old truck and is only reliable 70% of the time. Truck 2 is three years old and is reliable 90% of the time. Truck 3 is brand new and is reliable 95% of the time. What is the overall system reliability if the goal is to always have at least one pumper truck available?

(A) 0.598
(B) 0.700
(C) 0.950
(D) 0.998

18. Regression methods were used to analyze the data from a study investigating the relationship between compressive strength, x, and intrinsic permeability, y, of various concrete mixes and cures. Summary quantities are $n = 14$, $\sum y_i = 572$, $\sum y_i^2 = 23,530$, $\sum x_i = 43$, $\sum x_i^2 = 157.42$, and $\sum x_i y_i = 1697.80$. Use the equation of the fitted line to predict what permeability would be observed when the compressive strength is $x = 3.7$. If the observed value of permeability at $x = 3.7$ is $y = 46.1$, calculate the value of the corresponding residual.

(A) prediction = 39.39, residual = −6.61
(B) prediction = 39.39, residual = 6.61
(C) prediction = 48.01, residual = −1.91
(D) prediction = 48.01, residual = 1.91

19. The following pseudocode is a simulation program.

```
Number("WIN") = 0
Procedure Game
    Generate a uniform random number, U,
    within the range of 0 and 1;
    If U < 0.20, then mark = "WIN";
        Number("Win") = Number("WIN") + 1
    Otherwise mark = "LOSS";
End Procedure Game
Repeat Procedure Game
```

The initial 10 random numbers, U, generated are

0.63 0.18 0.95 0.47 0.20 0.79 0.32 0.44 0.82 0.26

Based on the initial ten random numbers generated, the Number("WIN") will be equal to (i). If the Procedure Game is repeated 10,000 times, the 99.9% probability for Number("WIN") will be in the range of (ii).

Which of the following represents the value (i) and the range of (ii)?

(A) (i) 2; (ii) $0 <$ Number("WIN") < 2000
(B) (i) 2; (ii) $1960 <$ Number("WIN") < 2000
(C) (i) 1; (ii) $1880 <$ Number("WIN") < 2120
(D) (i) 1; (ii) $1999 <$ Number("WIN") < 2001

20. The following chart is a 5 sec moving average plot of the simulated utilization of a resource versus run time. Resource utilization is a key decision variable in the experiment. The system under consideration is nonterminating and, therefore, it is desirable to discard statistics gathered during the startup transient.

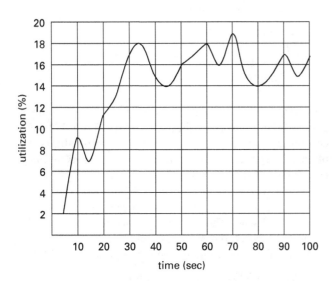

From the chart, which of the following is the most reasonable transient time for the experiment?

(A) 10 sec
(B) 20 sec
(C) 50 sec
(D) 100 sec

21. Which option presents potential solutions to the following problem?

$$\min f(x, y) = \frac{x^2}{2} - xy^2 + 4y^2$$

(A) $(x, y) = \{(0,0), (16,4)\}$
(B) $(x, y) = \{(0,0), (1,1)\}$
(C) $(x, y) = \{(0,0), (4,-2)\}$
(D) $(x, y) = \{(0,0), (9,-3)\}$

22. The following is a tableau of the simplex method for a maximization problem in standard form. Row Z represents the reduced cost for the problem. For this tableau, the values of a, b, and c that result in a unique optimal solution are

basic variable	eq. no.	x_1	x_2	x_3	x_4	x_5	x_6	x_7	RHS
Z	0	0	0	0	-8	-3	-2	a	
x_2	1	0	1	0	1	1	0	3	c
x_3	2	0	b	1	-2	2	-5	-1	2
x_1	3	1	0	0	0	-1	2	1	3

(A) $a = -4$, $b = 2$, $c = 6$
(B) $a = -2$, $b = 0$, $c = -4$
(C) $a = 0$, $b = 0$, $c = 3$
(D) $a = -2$, $b = 0$, $c = 8$

23. Customers arrive at a single-service facility at a Poisson distributed average rate of 40 per hour. When one customer is present, a single attendant operates the facility, and the service time for a customer is exponentially distributed with a mean value of 2 minutes. However, when there are two customers at the facility, the attendant is joined by an assistant and, working together, they reduce the mean service time to 1 minute. Assuming a system capacity of two customers, for what fraction of time are both servers idle?

(A) 0
(B) $^1/_4$
(C) $^1/_3$
(D) $^1/_2$

24. A bank employs two tellers to serve its customers. Customers arrive according to a Poisson process at a mean rate of 2 per minute. If a customer finds all tellers busy, he joins a queue that is served by all tellers. In other words, there are no lines in front of each teller, but rather one line waiting for the first available teller. The transaction time between the teller and customer has an exponential distribution with a mean rate of 1.5 per minute. What is the average number of customers in the bank?

(A) 2.4
(B) 2.7
(C) 3.0
(D) 3.6

25. Each record in a relational database table represents a unique combination of field values. Which type of entity relationship best captures the data item relationship between fields?

(A) $n{:}m$
(B) $n{:}1$
(C) $1{:}n$
(D) $1{:}1$

26. Job design is said to be management's most specific and detailed task in organizing the operations of an enterprise. Although there is no single "right" way to design jobs, most traditional management researchers agree that three key perspectives must be considered when designing a job. Which of the following is NOT a key job design perspective?

(A) span of control
(B) specialization
(C) job enrichment
(D) sociotechnical systems

27. In the following program evaluation and review technique (PERT) chart, what is the earliest time that event 4 can be realized?

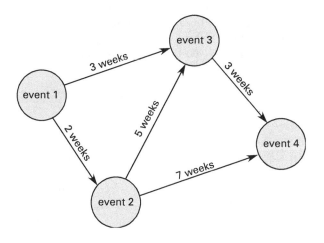

(A) 6 wk
(B) 9 wk
(C) 10 wk
(D) 20 wk

28. Near the end of each fiscal year, a production department manager sits down with his foreman and fills in the chart shown. (See *Chart for Problem 28.*)

The completed chart is used to determine the production supervisor's raise for that year. This job evaluation system could best be described as

(A) pay for results
(B) pay for merit
(C) cost-of-living pay
(D) seniority-based pay

29. A company uses a piecework incentive system based on standard hours with an 8 hr guarantee. The hourly rate is $7.50, and the standard is 3.0 hr per hundred finished parts assembled. How much would a worker who assembles 400 pieces in one day be paid?

(A) $60
(B) $75
(C) $90
(D) $120

30. Company ABC is a large international manufacturing firm with five divisions that are all organized as separate profit centers. All of the divisions use common products that are purchased from outside vendors. Several years ago, the company decided to standardize the purchasing of these items by using a central purchasing group. Recently, it has been recommended that the purchasing function be distributed back to the divisions. Which set of issues should NOT be considered when making this purchasing organization decision?

(A) cost, quality
(B) lead-time, expertise
(C) politics, past contracts
(D) volume, location

31. A project has four key activities: A_1, A_2, A_3, and A_4. The following table contains the precedence between activities, and the estimates of the optimistic, most likely, and pessimistic durations for each activity. Calculate the expected project duration and its variance.

Chart for Problem 28

	far exceeds job requirements	exceeds job requirements	meets job requirements	needs some improvement	does not meet minimum requirements
quality					
timeliness					
initiative					
adaptability					
communications					

		duration (hr)		
activity	precedes	optimistic	most likely	pessimistic
1-2	2-4	5	5	12
1-3	3-4	5	6	9
2-4	–	4	4	4
3-4	–	1	4	4

(A) $\mu = 9.00$, $\sigma^2 = 1.17$
(B) $\mu = 9.83$, $\sigma^2 = 0.69$
(C) $\mu = 10.00$, $\sigma^2 = 1.17$
(D) $\mu = 10.17$, $\sigma^2 = 1.36$

32. Amaprize, Inc., specializes in manufacturing party supplies. The cardboard required for the manufacture of birthday hats is ordered from Old West Paper Company. The cost to place one order is $50. The cardboard costs $0.40 per yd^2. The holding cost of inventory is 20% per year. The annual demand is 900,000 hats per year, and 1 ft^2 of cardboard is needed per hat. The economic order quantity for the cardboard is most nearly

(A) 1100 yd^2
(B) 7900 yd^2
(C) 11,000 yd^2
(D) 34,000 yd^2

33. The Watch Your Weight Fitness Company manufactures a variety of home fitness equipment. Its line of treadmills exhibits a seasonal demand pattern, with peaks in December and June.

quarter	sales forecast (units)
1	8000
2	12,000
3	6000
4	15,000

hiring cost = $100/employee
firing cost = $500/employee
beginning work force = 100 employees
production per employee = 100 units/quarter

Given the previous costs and quarterly sales forecasts, the cost of a chase (zero inventory) production strategy is most nearly

(A) $41,000
(B) $51,000
(C) $53,000
(D) $58,000

34. Which of the following is NOT an element of a just-in-time production system?

(A) quick setups
(B) kanban production system
(C) small lot sizes
(D) increased work-in-process inventory

35. Experimentation with a sintered carbide cutting tool at a 0.125 in depth of cut has shown the Taylor wear constant, C, to be 1744 and the n exponent to be 0.25. Which of the following most closely estimates the tool life at a cutting speed of 825 ft per minute?

(A) 0.20 min
(B) 1.2 min
(C) 20 min
(D) 200 min

36. A single-pass rough cut is performed on a face mill to reduce the thickness of a part. The mill is 3.5 in wide and has a depth-of-cut of 0.125 in. The workpiece will have a speed of 15 in per minute. The material removal rate for this operation is most nearly

(A) 0.44 in^3/min
(B) 4.2 in^3/min
(C) 6.6 in^3/min
(D) 13 in^3/min

37. One of the Design for Assembly (DFA) principles is to avoid nesting, tangling, or wedging. Based on this DFA principle, which statement best describes a redesign that will keep three parts with a trapezoidal cross section as shown from getting jammed up while moving on a conveyor track?

(A) Increase the wedge angle of the sloped sides.
(B) Add a lip around the bottom.
(C) Increase the height of the part while keeping the top and bottom dimensions constant.
(D) Any one of the above.

38. Part A takes 0.2 hr to make on machine M. Part B takes 0.1 hr to make on the same machine M. There are 8 hours per shift available to make parts A and B on machine M. How many M machines are required to meet a production rate of 60 pieces of part A and 120 pieces of part B in one shift?

(A) 2
(B) 3
(C) 4
(D) 5

39. Jobs J_1 through J_4 are processed on a single machine. Each job's due date is the number of days after processing begins (starting at $t = 0$). What job sequence minimizes the maximal tardiness (T_{max})?

job	process time	due date
J_1	2	10
J_2	3	3
J_3	4	6
J_4	5	8

(A) $\{J_1, J_2, J_3, J_4\}$, $T_{max} = 6$
(B) $\{J_2, J_1, J_3, J_4\}$, $T_{max} = 4$
(C) $\{J_2, J_3, J_4, J_1\}$, $T_{max} = 3$
(D) $\{J_4, J_3, J_2, J_1\}$, $T_{max} = 0$

40. If products can be grouped into families of similar products that can be produced by a group of workstations, which type of layout should be implemented?

(A) product layout
(B) process layout
(C) cellular layout
(D) fixed position layout

41. The production department of a company has three critical resources: three machine operators, four assemblers, and two inspectors. Each machine operator can produce 35 parts per hour, each assembler can assemble 25 parts per hour, and each inspector can inspect 60 parts per hour. The maximum production rate per 8 hour shift for the department is most nearly

(A) 800 parts
(B) 880 parts
(C) 960 parts
(D) 1000 parts

42. E-world.com is considering building a new distribution center to support its rapidly expanding e-commerce business. E-world's customer base is evenly distributed across the United States; however, it currently has suppliers located in the following cities only.

	x-y coordinates	% total product supplied
Los Angeles	$(200, 900)$	15%
Dallas	$(1500, 500)$	15%
Chicago	$(2100, 1500)$	30%
Miami	$(2800, 0)$	20%
Boston	$(3000, 1600)$	20%

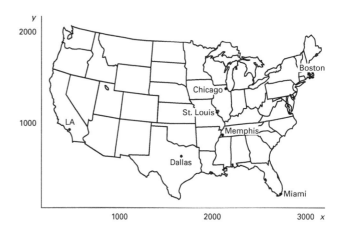

E-world.com wants to minimize the total distance that the suppliers are from the distribution center. Where should the new distribution center be located?

(A) anywhere on a line between $(2100, 900)$ Memphis and $(2100, 1500)$ Chicago
(B) at coordinate $(2100, 1500)$ Chicago
(C) anywhere on a line between $(1500, 500)$ Dallas and $(2100, 1500)$ Chicago
(D) at coordinate $(1950, 900)$ St. Louis

43. An industrial engineer is given an assignment of designing a material handling system under the mandate that its highest priority be to support the "systems" principle. Which of the following states the design issues that must be addressed?

(A) The system should be designed to minimize variety and customization in the methods and equipment used.
(B) The material flow multiplied by distance moved must be minimized.
(C) The material flow should be integrated from receiving to storage to production to shipping.
(D) The environmental impact and energy consumption of material handling alternatives must be analyzed.

44. A company must determine the type of material handling equipment best suited for moving product A, which has a demand of 100,000 units per year and a flow distance of 200 m per product. Based on general principles, the options have been narrowed to either a pallet jack, forklift, or automated guided vehicle (AGV). Using the following table, what is the total cost associated with each type of equipment, and which is the optimum selection?

equipment type	units/load	operating cost ($/m)	annualized capital equipment cost ($)
pallet jack	120	0.025	250
forklift	220	0.004	5000
AGV	180	0.001	15,000

(A) pallet jack = $4416, forklift = $5363,
      AGV = $15,111; select pallet jack

(B) pallet jack = $500,250, forklift = $85,000,
      AGV = $35,000; select AGV

(C) pallet jack = $833, forklift = $454,
      AGV = $555; select forklift

(D) pallet jack = $4416, forklift = $5363,
      AGV = $15,111; select AGV

45. It is most appropriate and economical to use a forklift-based material handling system when

(A) material flow rate is medium to high, distance is short to medium, and product routing is fixed

(B) material flow rate is low to medium, distance varies from short to medium, and product routing is variable but well defined

(C) material flow rate is low to medium, distance is highly variable, and product routing is highly variable

(D) material flow rate is high, distance is long, and product routing is moderately variable

46. Three principles of flow planning are

I. maximize directed flow (i.e., minimize cross flow and backtracking)

II. minimize flow (i.e., point of use delivery, combine flows and operations)

III. minimize cost of flow (i.e., reduce travel distances)

Which of the following material flow layouts best conforms to the previous three principles? (Note: The systematic layout planning (SLP) approach represents the strength of relationship or quantity of flow by the thickness of the arrow connecting two departments. Cell lines denote department boundaries.)

(A)

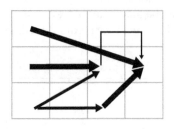

(B)

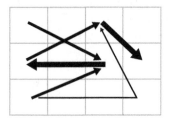

(C)

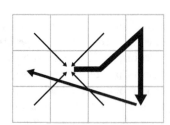

(D)

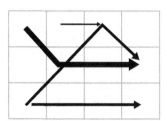

47. Castings are to be lifted from a pallet and placed on a grinding workstation that is 42 in high. The average vertical location above the floor where the hands begin the lifting operation is 20 in, and the average horizontal distance of the hands from the body's center of gravity at the beginning of the lift is 22 in. The maximum frequency that can be maintained for this task is 12 lifts per minute. The castings will then slide to the next workstation, and the job requires only one lift per cycle. The standard time for this job is 1.6 min per unit. The maximum permissible weight limit (in pounds per force) for this lifting task according to the NIOSH formula is most nearly

(A) 18 lbf
(B) 53 lbf
(C) 78 lbf
(D) 81 lbf

48. A cart is being designed to be pushed by an operator. The cart will be used to deliver material from the warehouse to the housewares section of a large department store. It is very important that the operator be able to see over the top of the cart to avoid customers who may be shopping in the area. Assuming an average shoe height of 2 cm, the maximum height of the cart should be approximately

(A) 140 cm
(B) 151 cm
(C) 153 cm
(D) 161 cm

49. A box weighing 40 lbf rests on a table. At the beginning of the lift, the horizontal distance of the hands grasping the box from the body's center of gravity is 28 in. The height of the table is 40 in. The overcoming moment required by the body to lift the package is most nearly

(A) 95 ft-lbf
(B) 100 ft-lbf
(C) 150 ft-lbf
(D) 200 ft-lbf

50. A time study consists of four elements, with element number 2 being a machine-control element. Element 4 occurs every fifth cycle. The operator is rated at 110% for all elements that are not machine controlled. Allowances for this task are 20%, and the following table shows the average observed time for each element.

element	average observed time (min)
1	0.14
2	2.12
3	0.16
4	5.28

What is most nearly the standard time for this task?

(A) 3.68 min
(B) 4.22 min
(C) 4.34 min
(D) 4.59 min

51. The following table summarizes the results of 200 observations to determine the delays in an operation.

observed activity	no. of times observed
lack of material	25
maintenance	20
quality check	5
machine operating	150

It is desired that the error for the delay activities be no larger than ±3% with a confidence of 95%. Approximately how many total observations are required of this task?

(A) 110
(B) 390
(C) 470
(D) 850

52. A job shop has an order for 20 units of part XYZ. If the first unit will take 10 hours to produce, and it is estimated that there is a 90% learning curve for this type of work, approximately how much time should be budgeted for the completion of the entire order?

(A) 50 hr
(B) 130 hr
(C) 150 hr
(D) 200 hr

53. Which of the following are examples of typical input/output ratios used by companies to measure their performance?

(A) profit/sales, sales/employee, capacity used/max capacity, profit/total investment
(B) profit/sales, inventory/advertising cost, orders/delivery, average pay/employee
(C) material cost/sales, employees/department, profit/total investment, defects/order
(D) employees/max capacity, sales/employee, sales/fixed assets, cost/unit

54. *Process capability* is defined as the ability of the process to meet design specifications. Every process has inherent variability, which can be measured. One measure of process capability is defined as

$$C_x = \frac{\text{USL}_x - \text{LSL}_x}{6\sigma_x}$$

C_x is the process capability measure, USL_x is the upper specifications limit as established by the part of process design, LSL_x is the lower specification limit, and $6\sigma_x$ is six standard deviations of the process producing the measured specification. It is assumed that the process producing the variable measure x is approximately normally distributed (i.e., comes from a Gaussian distribution).

The process is considered capable of meeting the designer's tolerances if

    (A) $C_x > 1$
    (B) $C_x \leq 1$
    (C) $C_x > 2$
    (D) $C_x = 6$

55. Given the definition of process capability described in Prob. 54, assume the process natural tolerance limits are equal to the process specification limits, and that no more than 3 parts in 1000 are expected to fall outside a specification limit. Which statement below is most nearly correct?

    (A)   The process mean is centered halfway between specification limits.
    (B)   The process mean is *not* centered halfway between specification limits.
    (C)   The centering of the process mean is not important in this calculation.
    (D)   The process is considered to be "six-sigma."

56. A process control chart on two parts—B, a bearing, and A, an axle—are found to be stable and normally distributed with mean μ_B, standard deviation σ_B, mean μ_A, and standard deviation σ_A, respectively.

If two parts are later randomly assembled, the clearance, X, between the axle and bearing is normally distributed. The probability that the clearance between the two parts is neither greater than 0.012 units, nor less than 0.005 units is

$$1 - \big(P(X_c < 0.005|\mu_B - \mu_z, \sigma_c)$$
$$+ P(X_c > 0.012|\mu_B - \mu_A, \sigma_c)\big)$$
$$= 1 - \left(P\left(z \leq \frac{0.005 - (\mu_B - \mu_A)}{\sigma_c}\right)\right.$$
$$\left. + P\left(z \geq \frac{0.012 - (\mu_B - \mu_A)}{\sigma_c}\right)\right)$$

$$X_c = \text{clearance} = X_B - X_A$$
$$\sigma_c = \text{standard deviation of clearance}$$

z is the standard normally distributed variable with mean of 0 and standard deviation of 1.

Select the proper formulation for the term σ_c in the denominator.

    (A) $\sigma_B + \sigma_A$
    (B) $\sigma_B - \sigma_A$
    (C) $\sqrt{\sigma_B^2 + \sigma_A^2}$
    (D) $\sqrt{|\sigma_B^2 - \sigma_A^2|}$

57. The Paper Machinery and Products Company sells office equipment and supplies to many large- and medium-sized companies. The company has received complaints concerning late delivery and setup of ordered equipment, but the company's records indicate that a high percentage ($> 95\%$) of machines are delivered on time to customers. The situation has deteriorated to the point where sales are being lost.

Which of the following seven tools of quality would be most appropriate to help the company solve its problem?

    I. going with the flow (chart)
    II. cause and effect diagrams
    III. control charts
    IV. histograms
    V. check sheets
    VI. Pareto charts
    VII. scatter diagrams

    (A)   II
    (B)   I and V
    (C)   III and IV
    (D)   VI and VII

58. The Pareto principle can be phrased as follows: 80% of the problems in any organization come from 20% of the causes. Select the answer that is most nearly correct.

    (A)   An organization should concentrate on the 80% of problems to reduce inefficiencies.
    (B)   An organization should concentrate on the 20% of problems to reduce inefficiencies.
    (C)   The employees of an organization can control only 20% of the problems; 80% is under the control of management.
    (D)   An organization should concentrate on 20% of the causes to reduce inefficiencies.

59. A company has implemented a single-sampling plan on its new XYZ product. The lot size is 3000 and the plan is to inspect 1%, or 30 units per lot. The lot will be accepted if at most 1 defect is found. What is most nearly the probability of accepting a lot given that the lot has a 3% defect rate?

(A) 0.40
(B) 0.77
(C) 0.83
(D) 0.97

60. A company has collected 10 samples, each with a sample size of 4, and determined the average, $\overline{X}_i$, and range, R_i.

sample i	$\overline{X}_i$	R_i
1	104	3
2	106	5
3	105	2
4	104	7
5	106	3
6	105	4
7	101	5
8	109	2
9	109	9
10	102	6

Calculate most nearly the 6σ spread for this process.

(A) 2.2
(B) 6.7
(C) 13
(D) 28

SOLUTIONS

1. In a benefit-cost analysis, in order for a project to be recommended, the present worth of the project's benefits should exceed the present worth of the project's costs. In other words, the ratio of benefits to costs needs to equal or exceed one.

benefit of elements $3,500,000 in year 1, increasing $50,000 each year thereafter (toll revenue)

cost elements $22,500,000 in year 0 (initial construction) and $475,000 each year thereafter (annual operating and maintenance costs)

To find the present worth of the benefit elements, use the uniform series present-worth factor in conjunction with the uniform gradient present-worth factor (to account for the uniform annual increase in revenue).

$$B = A(P/A, i, n) + G(P/G, i, n)$$

A = uniform series of end-of-compounding-period cash flows = $3,500,000
i = interest rate per compounding period = 12%/yr
n = number of compounding periods = 30 yr
G = uniform change from one period to the next in cash flows = $50,000

$$
\begin{aligned}
B &= \text{present worth of benefit elements} \\
&= (\$3{,}500{,}000)(P/A, 12\%, 30) \\
&\quad + (\$50{,}000)(P/G, 12\%, 30) \\
&= (\$3{,}500{,}000)\left(\frac{(1+0.12)^{30}-1}{(0.12)(1+0.12)^{30}}\right) \\
&\quad + (\$50{,}000)\left(\begin{array}{c}\dfrac{(1+0.12)^{30}-1}{(0.12)^2(1+0.12)^{30}} \\[2mm] -\dfrac{30}{(0.12)(1+0.12)^{30}}\end{array}\right) \\
&= (\$3{,}500{,}000)(8.0552) + (\$50{,}000)(58.7821) \\
&= \$31{,}132{,}305
\end{aligned}
$$

To find the total present worth of the cost elements, use the uniform series present-worth factor to find the present worth of the annual operating and maintenance costs, and add this value to the initial construction cost.

$$
\begin{aligned}
C &= \text{present worth of cost elements} \\
&= \$22{,}500{,}000 + A(P/A, i, n)
\end{aligned}
$$

$$A = \$475{,}000$$
$$i = 12\%/\text{yr}$$
$$n = 30 \text{ yr}$$

$$C = \$22{,}500{,}000 + (\$475{,}000)(P/A, 12\%, 30)$$
$$= \$22{,}500{,}000 + (\$475{,}000)\left(\frac{(1+0.12)^{30} - 1}{(0.12)(1+0.12)^{30}}\right)$$
$$= \$22{,}500{,}000 + (\$475{,}000)(8.0552)$$
$$= \$26{,}326{,}220$$

The benefit-cost ratio is

$$\frac{B}{C} = \frac{\$31{,}132{,}305}{\$26{,}326{,}220} = 1.18$$

Since the $B/C \geq 1$, the toll bridge should be built.

The answer is B.

2. The following equation is used to compute depreciation using the modified ACRS method.

$$D_j = (\text{factor})C$$
$$\text{factor} = \text{recovery rate for year } j$$
$$\qquad \text{(expressed as a decimal)}$$
$$\qquad = 0.20 \text{ for year 1 of a 5 yr recovery period}$$
$$\qquad = 0.32 \text{ for year 2 of a 5 yr recovery period}$$
$$C = \text{initial cost of the depreciable asset}$$
$$\qquad = \$250{,}000$$

The book value of an asset at the end of year j is equal to its initial cost less accumulated depreciation.

$$BV_j = C - \sum_{i=1}^{j} D_j$$

$$BV_2 = \text{book value at the end of year 2}$$
$$= C - D_1 - D_2$$
$$= \$250{,}000 - (0.20)(\$250{,}000) - (0.32)(\$250{,}000)$$
$$= \$120{,}000$$

The answer is B.

3. Capitalized costs are present worth values when the analysis period is an indefinite length of time. In general, capitalized cost $= A/i$, where A is a uniform series of end-of-period cash flows of indefinite length and i is the interest rate per compounding period.

In this problem, there is a one-time cash flow of $120,000, an infinite end-of-year series of $2000, and a uniform series of end-of-year cash flow of $6500 − $2000 = $4500 in years 1 through 20.

capitalized
$$\text{cost} = \$120{,}000 + A_1(P/A, i, n) + A_2/i$$
$$A_1 = \text{finite uniform series of cash flows}$$
$$\qquad = \$4500$$
$$i = \text{interest rate per compounding period}$$
$$\qquad = 10\%/\text{yr}$$

$$n = \text{number of compounding periods}$$
$$= 20 \text{ yr}$$
$$A_2 = \text{uniform series of cash flows of}$$
$$\qquad \text{indefinite length}$$
$$= \$2000$$

capitalized
$$\text{cost} = \$120{,}000 + (\$4500)(P/A, 10\%, 20)$$
$$+ \frac{\$2000}{0.10}$$
$$= \$120{,}000 + (\$4500)\left(\frac{(1+0.10)^{20} - 1}{(0.10)(1+0.10)^{20}}\right)$$
$$+ \frac{\$2000}{0.10}$$
$$= \$178{,}311 \quad (\$170{,}000)$$

The answer is A.

4. Overhead costs typically consist of all costs that are not direct labor or direct material costs. Since raw materials are direct material costs, they are not factored into overhead.

The answer is C.

5. To determine overhead cost, first find the overhead rate.

$$\text{overhead rate} = \frac{\text{total overhead in dollars for period}}{\text{total machine hours for period}}$$
$$= \frac{\$200{,}000}{18{,}000 \text{ hr}}$$
$$= \$11.11/\text{hr}$$

$$\frac{\text{overhead cost}}{\text{unit}} = (\text{overhead rate})\left(\frac{\text{machine hours}}{\text{unit}}\right)$$
$$= \left(11.11 \, \frac{\$}{\text{hr}}\right)\left(0.75 \, \frac{\text{hr}}{\text{unit}}\right)$$
$$= \$8.33/\text{unit}$$

The answer is A.

6. Let X be the number of tricycles sold to make a profit of $10,000 per year.

$$\text{profit} = \text{revenues} - \text{costs} = \$10{,}000/\text{yr}$$
$$\text{revenues} = \left(\frac{\text{selling price}}{\text{unit}}\right)\left(\begin{array}{c}\text{number of}\\\text{units sold}\end{array}\right)$$
$$= \left(40 \, \frac{\$}{\text{tricycle}}\right)\left(X \, \frac{\text{tricycles}}{\text{yr}}\right)$$
$$\text{costs} = \text{fixed cost} + \text{variable cost}$$
$$\text{fixed cost} = \$8000/\text{yr}$$
$$\text{variable cost} = \left(20 \, \frac{\$}{\text{tricycle}}\right)\left(X \, \frac{\text{tricycles}}{\text{yr}}\right)$$

$$10,000 \; \frac{\$}{\text{yr}} = \left(40 \; \frac{\$}{\text{tricycle}}\right)\left(X \; \frac{\text{tricycles}}{\text{yr}}\right)$$
$$- 8000 \; \frac{\$}{\text{yr}}$$
$$- \left(20 \; \frac{\$}{\text{tricycle}}\right)\left(X \; \frac{\text{tricycles}}{\text{yr}}\right)$$
$$18,000 \; \frac{\$}{\text{yr}} = \left(20 \; \frac{\$}{\text{tricycle}}\right)\left(X \; \frac{\text{tricycles}}{\text{yr}}\right)$$
$$X = 900 \; \text{tricycles/yr}$$

The answer is D.

7. Estimating the cost of the new facility requires calculating two components—the impact of construction cost index and the impact of size increase. Both are a ratio based on the change of their respective factors, with the size increase modified by the power law sizing exponent. The equation for this estimate is as follows.

$$C_{\text{new}} = C_{\text{old}}\left(\frac{I_{\text{new}}}{I_{\text{old}}}\right)\left(\frac{Q_{\text{new}}}{Q_{\text{old}}}\right)^{e}$$

C_{old} = cost of old asset = \$350 million
I_{old} = construction cost index of old asset = 121
I_{new} = construction cost index of new asset = 139
Q_{old} = size of old asset = 400 MW
Q_{new} = size of new asset = 1000 MW
e = power law exponent = 0.75

$$C_{\text{new}} = (\$350 \; \text{million})\left(\frac{139}{121}\right)\left(\frac{1000 \; \text{MW}}{400 \; \text{MW}}\right)^{0.75}$$
$$= \$799.37 \; \text{million} \quad (\$800 \; \text{million})$$

The answer is C.

8. The straight line depreciation is calculated as

$$D = \frac{C - S_n}{n}$$

C = initial cost = \$140,000
S_n = salvage value = \$20,000
n = depreciation life = 7 yr

$$D = \frac{\$140,000 - \$20,000}{7 \; \text{yr}}$$
$$= \$17,143/\text{yr}$$

The book value at the end of year 4 is

$$BV_t = C - tD$$
$$BV_4 = \$140,000 - (4 \; \text{yr})\left(17,143 \; \frac{\$}{\text{yr}}\right)$$
$$= \$71,428 \quad (\$71,000)$$

The answer is D.

9. The equation for after-tax cash flow (ATCF) is

$$\text{ATCF}_n = (1-t)(R_n - C_n - \text{IP}_n) + tD_n - \text{PP}_n$$

R_n = revenue at time n = \$800,000
C_n = cost (i.e., all expenses) at time n
     = \$300,000
D_n = depreciation at time n = \$100,000
IP_n = interest payment at time n = \$40,000
PP_n = principle payment at time n
     = \$80,000 − \$40,000 = \$40,000
t = tax rate = 0.40

$$\text{ATCF}_n = (1\text{-}0.40)(\$800,000 \text{ - } \$300,000 \text{ - } \$40,000)$$
$$+ (0.40)(\$100,000) \text{ - } \$40,000$$
$$= \$276,000 + \$0$$
$$= \$276,000$$

The answer is B.

10. Use the NCEES Handbook's one-way ANOVA equations to determine if there is an effect due to changing the sand mix. (One-way tests are appropriate when there are three or more independent tests.)

n = sample size = 4
N = total number of observations = 16
T = total of all N observed values = 126.71
α = significance level = 0.05
df_1 = degrees of freedam 1 = 4 − 1 = 3
df_2 = $N - n$ = 12

$$\text{SS} = \sum_{i=1}^{k}\sum_{j=1}^{n_i}(x_{ij} - \overline{x})^2$$

$$C = T^2/N$$

$$\text{SS}_{\text{total}} = \sum_{i=1}^{k}\sum_{j=1}^{n_i}x_{ij}^2 - C$$

$$\text{SS}_{\text{treatments}} = \sum_{i=1}^{k}\left(\frac{T_i^2}{n_i}\right) - C$$

$$\text{SS}_{\text{error}} = \text{SS}_{\text{total}} - \text{SS}_{\text{treatments}}$$

The sum of squares (SS) is

$$\text{SS} = (7.34)^2 + (8.41)^2 + \ldots + (6.4)^2 = 1017.44$$

The total sum-of-square errors is

$$\text{SS}_{\text{total}} = 1017.44 - \frac{(126.71)^2}{16} = 13.98$$

The sum-of-square errors between samples (treatments) is

$$\text{SS}_{\text{treatments}} = \frac{(31.38)^2}{4} + \frac{(29.13)^2}{4} + \frac{(36.66)^2}{4}$$
$$+ \frac{(29.54)^2}{4} - \frac{(126.71)^2}{16}$$
$$= 8.993$$

The error sum of squares is

$$SS_{error} = SS_{total} - SS_{treatments}$$
$$= 13.98 - 8.99$$
$$= 4.988$$

The mean square error between treatments (MST) is

$$MST = \frac{SS_{treatments}}{4 - 1} = \frac{8.993}{4 - 1}$$
$$= 2.998$$

The mean square error (MSE) is

$$MSE = \frac{SS_{error}}{16 - 4} = \frac{4.988}{12}$$
$$= 0.416$$

$$F = \frac{MST}{MSE} = \frac{2.998}{0.416}$$
$$= 7.21$$

7.21 is greater than $F(0.5, 3, 12) = 3.49$ (from the "Critical Values of the F-Distribution" table in the NCEES Handbook).

The sand mix has a significant effect on the compression resistance of structure supports.

The answer is D.

11. Use linear regression.

$$S_{xy} = \sum_{i=1}^{n} x_i y_i - \left(\frac{1}{n}\right) \left(\sum_{i=1}^{n} x_i\right) \left(\sum_{i=1}^{n} y_i\right)$$
$$= 1{,}018{,}790 - \left(\tfrac{1}{6}\right)(3300)(1876)$$
$$= -13{,}010$$

$$S_{xx} = \sum_{i=1}^{n} x_i^2 - \left(\frac{1}{n}\right) \left(\sum_{i=1}^{n} x_i\right)^2$$
$$= 1{,}828{,}450 - \left(\tfrac{1}{6}\right)(3300)^2$$
$$= 13{,}450$$

$$b = \frac{S_{xy}}{S_{xx}}$$
$$= -0.967$$

$$\overline{y} = \left(\frac{1}{n}\right) \left(\sum_{i=1}^{n} y_i\right)$$
$$= 312.67$$

$$\overline{x} = \left(\frac{1}{n}\right) \left(\sum_{i=1}^{n} x_i\right)$$
$$= 550$$

Then,

$$a = \overline{y} - b\overline{x}$$
$$= 312.67 - (-0.967)(550)$$
$$= 844.52$$

The regression line is $y = 844.52 - 0.967x$.

The answer is B.

12. Experimental runs must be randomized throughout the day, preferably with repetition in different time slots. These experiments were conducted sequentially with incremental factor measurements; thus, additives may be confounded with the rising temperature during the day. In this case, results from a statistical analysis will most likely provide no meaningful conclusion.

The answer is D.

13. A 90% confidence interval on mean entity time-in-system says there is a 90% chance that the population (or true) mean time-in-system lies within the given interval. Therefore, choice (A) is incorrect. Further, such a confidence interval says nothing either about the individual data values that were used to construct the interval or about future individual experimental values. Therefore, choices (B) and (C), respectively, are incorrect. Because a replicate approach was used in the experiment, the central limit theorem applies, and it can be assumed that the collected data follows a normal distribution. Since the data is normally distributed, it is reasonable to assume that the 10% error (100% − 90%) is equally distributed above and below the confidence interval.

The answer is D.

14. Let $\overline{x}$ represent the sample mean of a random sample of size n from a population with known variance σ^2. A $100(1 - \alpha)\%$ confidence interval on the true mean μ using $z_{\alpha/2}$, the upper $100\alpha/2$ percentage point of the standard normal distribution, is given by

$$\overline{x} - z_{\alpha/2} \left(\frac{\sigma}{\sqrt{n}}\right) \leq \mu \leq \overline{x} + z_{\alpha/2} \left(\frac{\sigma}{\sqrt{n}}\right)$$

The error in estimating the mean life is equal to $\mu - \overline{x}$.

$$\mu - \overline{x} = z_{\alpha/2} \left(\frac{\sigma}{\sqrt{n}}\right)$$

$$n = \left(\frac{z_{\alpha/2}\sigma}{\mu - \overline{x}}\right)^2$$
$$= \left(\frac{(1.96)(25)}{5}\right)^2$$
$$= 96.04 \quad (97)$$

The answer is D.

15. Since each person has a fixed probability of 0.95 of showing up, the probability that N passengers show up out of 52 follows a binomial distribution with (52, 0.95). In order to accommodate all passengers, N should be less than or equal to 50.

$$P(N \leq 50) = 1 - P(N = 51) - P(N = 52)$$
$$= 1 - \binom{52}{51}(0.95)^{51}(0.05)^1$$
$$- \binom{52}{52}(0.95)^{52}(0.05)^0$$
$$= 0.74$$

The answer is B.

16. With 13 items, the median value will be item 7 in the ordered list. List the data in increasing order.

3.05, 3.11, 3.12, 3.12, 3.15, 3.15, 3.16, 3.16, 3.17, 3.18, 3.19, 3.19, 3.20

Counting to the 7th data item, the median is 3.16. The mean value is

$$\overline{x} = \frac{\sum x_i}{n} = 3.15$$

Calculate the sample standard deviation.

$$s = \sqrt{\frac{\sum\limits_{i=1}^{n}(x_i - \overline{x})^2}{n - 1}}$$

$$= \sqrt{\frac{(3.05 - 3.15)^2 + (3.11 - 3.15)^2 + \ldots + (3.20 - 3.15)^2}{13 - 1}}$$

$$= 0.0414$$

The answer is D.

17. This is a parallel system reliability problem since only one truck needs to be working to provide fire protection service. For n independent components connected in parallel, the probability that the system will function is

$$R(P_1, P_2, \ldots P_n) = 1 - \prod_{i-1}^{n}(1 - P_i)$$

P_i is the probability that component i is functioning.

Therefore, the overall system reliability is

$$R(0.70, 0.90, 0.95) = 1 - (1 - 0.70)(1 - 0.90)(1 - 0.95)$$
$$= 0.998$$

The answer is D.

18. The least square estimates of the slope interpreted in the simple linear regression model are

$$b = \frac{\sum\limits_{i=1}^{n} y_i x_i - \left(\frac{1}{n}\right)\left(\sum\limits_{i=1}^{n} y_i\right)\left(\sum\limits_{i=1}^{n} x_i\right)}{\sum\limits_{i=1}^{n} x_i^2 - \left(\frac{1}{n}\right)\left(\sum\limits_{i=1}^{n} x_i\right)^2}$$

$$= \frac{1697.80 - \dfrac{(572)(43)}{14}}{157.42 - \dfrac{(43)^2}{14}}$$

$$= -2.33$$

$$a = \frac{1}{n}\sum_{i=1}^{n} y_i - \frac{b}{n}\sum_{i=1}^{n} x_i = \frac{572}{14} + \frac{(2.33)(43)}{14}$$

$$= 48.01$$

The fitted or estimated regression value is

$$y = a + bx = 48.01 - (2.33)(3.7) = 39.39$$

The residual value is

$$y_{\text{actual}} - y = 46 - 39.39 = 6.61$$

The answer is B.

19. (i) Among the ten random numbers, there is only one number that is smaller than 0.20 (0.18 < 0.20). Thus, there will be only one "WIN" in the simulated game.

(ii) When the sample is as large as 10,000, the binomial distribution ($p = 0.2$, $q = 0.8$) can be approximated by a normal distribution with $N(\mu = np, \sigma = \sqrt{npq})$.

$$n = 10{,}000$$
$$\mu = np = (10{,}000)(0.2) = 2000$$
$$\sigma = \sqrt{npq} = \sqrt{(10{,}000)(0.2)(0.8)} = 40$$

The 99.9% probability limits are $\pm 3\sigma$ range from the mean; that is, $2000 \pm 3(40)$, or 1880 and 2120.

The answer is C.

20. When discarding transient data, it is desirable to discard enough data so as not to bias the statistics with data from the start-up phase, while at the same time not to discard so much data as to become inefficient. Conservative rules-of-thumb suggest erring on the side of discarding too much versus too little data. In this problem, at the points suggested by 10 sec and 20 sec, the system has clearly not yet stabilized. If the point suggested by 100 sec were selected, too much would be discarded. Therefore, 50 sec appears to be a good transient cutoff point.

The answer is C.

21. This is an unconstrained optimization problem that can be solved using derivatives. Any solution to this problem must satisfy the first-order necessary condition, which is that the gradient of the function is equal to zero. That is, $\nabla f(x, y) = 0$.

$$\nabla f(x, y) = \begin{cases} x - y^2 = 0 \\ 8y - 2xy = 0 \end{cases}$$

To solve this set of two equations and two unknowns, solve the first equation for x to yield $x = y^2$. Substitute y^2 for x into the second equation to yield $8y - 2y^3 = 0$. The solutions of this equation are $y = 0$, $y = 2$, and $y = -2$. Substituting back into the first equation gives the three solutions $(x, y) = \{(0, 0), (4, 2), (4, -2)\}$. Therefore, only choice (C) can be the solution. Since the $\nabla f(x, y) = 0$ is a necessary condition, an alternate approach would be to substitute each of the answers into the equations and see which ones satisfy the necessary conditions.

The answer is C.

22. In order that a solution to a maximization problem be unique and optimal when using the simplex method, the reduced costs must be negative for each of the non-basic feasible solutions. This implies that if any nonbasic feasible solution comes into the basis, the objective function will decrease in value and the current solution must be optimal. Likewise, to be a feasible solution to a problem in standard form, the solution must be non-negative. If the current right-hand side in a tableau contains a negative number, the solution is not feasible and a mistake must have been made. Also, in the tableau, the columns corresponding to the basic feasible solutions must form an identity matrix. Each column of a basic feasible variable has a 1 in the row of the matching basic variable and zeros elsewhere. For the variables listed, the following must hold: $a < 0$, $b = 0$, and $c \geq 0$.

The answer is D.

23. This problem can be viewed as an example of a continuous-time Markov chain, since both interarrival and service times are exponentially distributed. The states of this chain are 0, 1, and 2, which represent the number of customers in the system. Because the arrival rate is 40 per hour and the service rates are 30 or 60 per hour when there are 1 or 2 customers, respectively, the transition diagram is

To find the proportion of time that both servers are free, compute the steady-state probability of having no customers. This is done through solving the balance equations given by

$$40\pi_0 = 30\pi_1$$
$$70\pi_1 = 40\pi_0 + 60\pi_2$$
$$60\pi_2 = 40\pi_1$$
$$\pi_0 + \pi_1 + \pi_2 = 1$$

This system of equations reduces to

$$\pi_1 = \tfrac{4}{3}\pi_0$$
$$\pi_2 = \tfrac{2}{3}\pi_0$$

From the normalizing equation,

$$\pi_0 + \tfrac{4}{3}\pi_0 + \tfrac{2}{3}\pi_0 = 1$$
$$\tfrac{9}{3}\pi_0 = 1$$
$$\pi_0 = 1/3$$

The proportion of time in which both servers are free is $^1/_3$.

The answer is C.

24. This problem follows an M/M/2 model. The interarrival and service times are exponentially distributed. The average number of customers in the bank is given by L. The arrival rate, λ, is 2 per minute, the service rate, μ, is 1.5 per minute, and the number of servers, s, is 2.

$$L = L_q + \frac{\lambda}{\mu}$$

$$L_q = \frac{P_0 s^s \rho^{s+1}}{s!(1 - \rho)^2}$$

The utilization is

$$\rho = \frac{\lambda}{s\mu} = \frac{2}{(2)(1.5)} = 2/3$$

$$s\rho = (2)\left(\tfrac{2}{3}\right) = 4/3$$

$$P_0 = \frac{1}{\displaystyle\sum_{n=0}^{s-1} \frac{(s\rho)^n}{n!} + \frac{(s\rho)^s}{s!(1 - \rho)}}$$

$$= \frac{1}{\dfrac{\left(\dfrac{4}{3}\right)^0}{0!} + \dfrac{\left(\dfrac{4}{3}\right)^1}{1!} + \dfrac{\left(\dfrac{4}{3}\right)^2}{2!\left(1 - \dfrac{2}{3}\right)}}$$

$$= 1/5$$

$$L_q = \frac{\left(\dfrac{1}{5}\right)(2)^2 \left(\dfrac{2}{3}\right)^3}{2!\left(1 - \dfrac{2}{3}\right)^2}$$

$$= 1.07$$

$$L = 1.07 + 1.33$$
$$= 2.4$$

The average number of customers in the bank is 2.4.

The answer is A.

25. For the records in a relational database table to exhibit a unique combination of data items, the fields must exhibit a 1:1 relationship. Other relationships are represented between tables with the help of unique key fields.

The answer is D.

26. Specialization (splitting a task into more specialized subtasks to improve productivity), job enrichment (expanding a task, both horizontally and vertically, to achieve greater enthusiasm and motivation), and sociotechnical systems (designing a task to ensure an appropriate blend of technology and human interaction in the workplace) are three widely recognized job design perspectives. Span of control can be defined as the number, or sometimes the limit on the number, of people that a given manager can supervise. As such, span of control is a perspective of organizational design, not of job design.

The answer is A.

27. The earliest time that event 4 can be realized is after the completion of the longest-duration sequence of activities that results in event 4. In this problem, activities 1-2, 2-3, and 3-4 create a total duration of 10 weeks, which is the longest-duration sequence.

The answer is C.

28. Paying for merit is a performance appraisal system that consists of lists of factors or traits divided into degrees upon which an individual can be assessed. For each factor or trait, the individual (often through interactive discussion through a manager) is assigned a performance degree. The resultant assessment is then used by the manager to determine pay raises.

The answer is B.

29. To determine how much the worker is paid, use the following equation.

$$\text{earned time} = \left(\begin{array}{c}\text{no. of parts}\\\text{assembled}\end{array}\right)\left(\frac{\text{standard time}}{100 \text{ parts}}\right)$$
$$= (400)\left(\frac{3.0 \text{ hr}}{100}\right)$$
$$= 12.0 \text{ standard hr}$$

Since more than eight hours of work were completed, the incentive system is used.

$$\text{pay} = (\text{earned hours})\left(\text{rate } \frac{\$}{\text{hr}}\right)$$
$$= (12.0 \text{ hr})\left(7.50 \frac{\$}{\text{hr}}\right)$$
$$= \$90$$

The answer is C.

30. Purchasing decisions should consider all system-wide factors that would affect overall company performance. Included among these factors are vendor lead time, employee expertise, overall product cost, overall product quality, product volume requirements, and vendor location. Internal company politics should never enter into the decision making process, and past contractual relationships should not limit future decisions.

The answer is C.

31. The network is

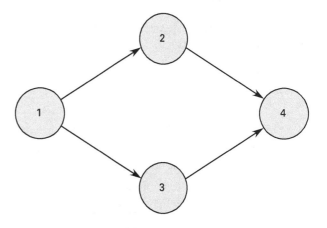

Use the project evaluation and review technique (PERT) mean and variance equations.

$$a_{ij}, b_{ij}, c_{ij} = \text{optimistic, most likely, pessimistic durations for activity } (i, j)$$
$$\mu_{ij} = \text{mean duration of activity } (i, j)$$
$$= \frac{a_{ij} + 4b_{ij} + c_{ij}}{6}$$
$$\sigma_{ij}^2 = \text{variance of the duration of activity } (i, j)$$
$$= \left(\frac{c_{ij} - a_{ij}}{6}\right)^2$$

Calculate each activity's mean and variance as given in the following table.

activity	a_{ij}	b_{ij}	c_{ij}	mean	variance
1-2	5	5	12	6.17	1.36
1-3	5	6	9	6.33	0.44
2-4	4	4	4	4.00	0.00
3-4	1	4	4	3.50	0.25

There are two paths through the network with the following expected times and variances.

$$\text{path 1-2-4 time} = 6.17 + 4.00 = 10.17$$
$$\text{variance} = 1.36 + 0.00 = 1.36$$

$$\text{path 1-3-4 time} = 6.33 + 3.50 = 9.83$$
$$\text{variance} = 0.44 + 0.25 = 0.69$$

The critical path is the longer of these two path times, or path 1-2-4, so $\mu = 10.17$ and $\sigma^2 = 1.36$.

The answer is D.

32. The economic order quantity (EOQ) is computed using the following equation.

$$\text{EOQ} = \sqrt{\frac{2AD}{rC}}$$

A = cost to place one order = \$50
D = number of units used per year
r = interest rate on inventory = 20%/yr
C = purchase price per unit = \$0.40/yd^2

Defining a unit to be 1 yd^2 of cardboard,

$$D = \left(900,000 \ \frac{\text{hats}}{\text{yr}}\right)\left(1 \ \frac{\text{ft}^2}{\text{hat}}\right)\left(\frac{1 \ \text{yd}^2}{9 \ \text{ft}^2}\right)$$
$$= 100,000 \ \text{yd}^2/\text{yr}$$

$$\text{EOQ} = \sqrt{\frac{(2)(\$50)\left(100,000 \ \dfrac{\text{yd}^2}{\text{yr}}\right)}{\left(\dfrac{0.20}{\text{yr}}\right)\left(0.40 \ \dfrac{\$}{\text{yd}^2}\right)}}$$
$$= 11,180 \ \text{yd}^2 \quad (11,000 \ \text{yd}^2)$$

The answer is C.

33. A chase (zero inventory) production strategy meets demand with minimal inventory investment by hiring and firing employees each period. The cost of this strategy consists of the cost of hiring and firing employees.

The forecasts of demand and employee production rate are used to compute the number of employees needed in each period. (See *Table for Solution 33*.)

$$\frac{\text{number of}}{\text{employees needed}} = \frac{\text{sales forecast}}{\text{production per employee}}$$

The number of employees hired/fired is determined by comparing the current workforce to the number of employees needed.

W_t = number of employees needed in period t
     (W_0 = beginning workforce level = 100)
H_t = number of employees hired at the beginning
     of period t = $\max\{W_t - W_{t-1}, \ 0\}$
F_t = number of employees fired at the beginning
     of period t = $\max\{W_{t-1} - W_t, \ 0\}$

Table for Solution 33

period, t	sales forecast	W_t	H_t	F_t
1	8000	$\dfrac{8000}{100} = 80$	$\max\{0, 80 - 1000\} = 0$	$\max\{100 - 80, 0\} = 20$
2	12,000	$\dfrac{12,000}{100} = 120$	$\max\{0, 120 - 80\} = 40$	$\max\{80 - 120, 0\} = 0$
3	6000	$\dfrac{6000}{100} = 60$	$\max\{0, 60 - 120\} = 0$	$\max\{120 - 60, 0\} = 60$
4	15,000	$\dfrac{15,000}{100} = 150$	$\max\{0, 150 - 60\} = 90$	$\max\{60 - 150, 0\} = 0$

The cost of this strategy is estimated by multiplying the cost of hiring an employee by the total number of employees hired over the planning horizon and adding this to the cost of firing an employee multiplied by the total number of employees fired.

$$\text{cost of chase strategy} = \left(100 \; \frac{\$}{\text{employee}}\right)(130 \text{ employees})$$
$$+ \left(500 \; \frac{\$}{\text{employee}}\right)(80 \text{ employees})$$
$$= \$53{,}000$$

The answer is C.

34. The elimination of waste is the basic tenet of a just-in-time (JIT) production system. Choices (A), (B), and (C) all strive to eliminate waste. Inventory is a primary example of waste, thus choice (D) is not an element of a JIT production system.

The answer is D.

35. Solve for T, the tool's lifespan. (The equation is not dimensionally consistent.)

$$vT^n = C$$
$$\left(825 \; \frac{\text{ft}}{\text{min}}\right) T^{0.25} = 1744$$
$$T^{0.25} = \frac{1744}{825 \; \dfrac{\text{ft}}{\text{min}}} = 2.114$$
$$T = (2.114)^{1/0.25}$$
$$= 19.96 \text{ min} \quad (20 \text{ min})$$

The answer is C.

36. From the problem statement, the material removal rate (MRR) is

$$\text{MRR} = (\text{width})(\text{depth of cut})(\text{workpiece speed})$$
$$= (3.5 \text{ in})(0.125 \text{ in})\left(15 \; \frac{\text{in}}{\text{min}}\right)$$
$$= 6.56 \text{ in}^3/\text{min} \quad (6.6 \text{ in}^3/\text{min})$$

The answer is C.

37. Any one of the parts shown is likely to slide up the sloped side of its neighboring part when pushed on a conveyor track, causing jamming. This outcome is caused by the narrow wedge or sloped side of the part. Increasing the wedge angle of the sloped side, as in choice (A), or adding a lip around the bottom, as in choice (B), will prevent the parts from jamming. Increasing the height of the part while keeping the top and bottom constant, as in choice (C), will also increase the wedge angle. So, any of the above redesigns will keep the parts from getting jammed while moving on a conveyor track.

The answer is D.

38. Use the following machine requirement equation to determine the number of machines, M_j, required to meet the desired production rate of parts A and B in one shift. In one shift, P_i is the desired production rate for part i, T_i is the production time for product i, and C_i is the number of hours available to make part i.

$$M_j = \sum_{i=1}^{n} \frac{P_{ij}T_{ij}}{C_{ij}}$$
$$= \frac{\left(\dfrac{60 \text{ part A}}{\text{shift}}\right)\left(\dfrac{0.2 \text{ hr}}{\text{part A}}\right)}{8 \; \dfrac{\text{hr}}{\text{shift}}}$$
$$+ \frac{\left(\dfrac{120 \text{ part B}}{\text{shift}}\right)\left(\dfrac{0.1 \text{ hr}}{\text{part B}}\right)}{8 \; \dfrac{\text{hr}}{\text{shift}}}$$
$$= 3$$

The answer is B.

39. When jobs are sequenced in a non-decreasing order by their due date (i.e., the earliest due date), the maximal tardiness is minimized on a single machine.

job	due date	completion time	lateness
J_2	3	$0 + 3 = 3$	0
J_3	6	$3 + 4 = 7$	1
J_4	8	$7 + 5 = 12$	4
J_1	10	$12 + 2 = 14$	4

(completion time = completion time of previous job + processing time of current job)

The sequence that minimizes the maximal tardiness is $\{J_2, J_3, J_4, J_1\}$ with $T_{\max} = 4$.

The answer is C.

40. Cellular layouts are appropriate when a product can be grouped into families of similar products that can be produced by a group of workstations. Product layouts are appropriate when the product is fairly standard and has a large, stable demand. Process layouts are appropriate when there is a high degree of variability in the type of products produced, with each product having a low demand. A fixed-position layout is appropriate when the product is large and hard to move and has a relatively low demand.

The answer is C.

41. In order to determine the bottleneck resource, the hourly rate of each resource must be determined.

$$\frac{\text{parts}}{\text{hr}} = (\text{no. of resources})\left(\frac{\text{production rate}}{\text{hr}}\right)$$

$$\text{machine operations} = (3)\left(35 \ \frac{\text{parts}}{\text{hr}}\right)$$
$$= 105 \ \text{parts/hr}$$

$$\text{assembly} = (4)\left(25 \ \frac{\text{parts}}{\text{hr}}\right)$$
$$= 100 \ \text{parts/hr}$$

$$\text{inspection} = (2)\left(60 \ \frac{\text{parts}}{\text{hr}}\right)$$
$$= 120 \ \text{parts/hr}$$

The bottleneck operation is assembly, limited to 100 parts per hour, resulting in a maximum production for 8 hours of

$$(8 \ \text{hr})\left(100 \ \frac{\text{parts}}{\text{hr}}\right) = 800 \ \text{parts}$$

The answer is A.

42. The optimal value for the location of a single facility in order to minimize the sum of the total interaction can be found by solving for each of the two coordinates (x and y) separately using two mathematical properties of the optimal solution. First, the x and y coordinates of the new facility will be one of the existing x and y coordinates of the existing facilities. Second, the optimal x and y will be located so that no more than half of the total weight is to the left of x and no more than half to the right of x.

Solving for the x coordinate,

	x	% total product supplied
Los Angeles	200	15%
Dallas	1500	15%
Chicago	2100	30% cum. > 50%; therefore, $x^* = 2100$
Miami	2800	20%
Boston	3000	20%

Solving for the y coordinate,

	y	% total product supplied
Miami	0	20%
Dallas	500	15%
Los Angeles	900	15% cum. = 50%; therefore, $y^* = 900$ or 1500
Chicago	1500	30%
Boston	1600	20%

The new distribution center can be located anywhere on a line between $(2100, 900)$ [approximately Memphis] and $(2100, 1500)$ [Chicago].

The answer is A.

43. To support the systems principle, the material flow should be integrated from receiving to storage to production to shipping. Choice (A) represents the standardization principle, choice (B) represents the work principle, and choice (D) represents the environmental principle.

The answer is C.

44. Calculate the total cost associated with each type of equipment.

$$\text{total cost} = (\text{operating cost})\left(\frac{\text{total demand}}{\frac{\text{units}}{\text{load}}}\right)$$
$$\times (\text{flow distance})$$
$$+ \text{annualized capital equipment cost}$$

For the pallet jack,

$$\text{total cost} = \left(0.025 \ \frac{\$}{\text{m}}\right)\left(\frac{100{,}000 \ \text{units}}{120 \ \frac{\text{units}}{\text{load}}}\right)(200 \ \text{m})$$
$$+ \$250$$
$$= \$4416$$

For the forklift,

$$\text{total cost} = \left(0.004 \ \frac{\$}{\text{m}}\right)\left(\frac{100{,}000 \ \text{units}}{220 \ \frac{\text{units}}{\text{load}}}\right)(200 \ \text{m})$$
$$+ \$5000$$
$$= \$5363$$

For the AGV,

$$\text{total cost} = \left(0.001 \ \frac{\$}{\text{m}}\right) \left(\frac{100{,}000 \text{ units}}{180 \ \frac{\text{units}}{\text{load}}}\right) (200 \text{ m})$$
$$+ \$15{,}000$$
$$= \$15{,}111$$

The pallet jack is the optimum selection with the smallest total cost of \$4416.

The answer is A.

45. Forklifts are most economical when used for highly variable production environments, and where flexibility is a desired characteristic of the material handling system. For instance, forklifts are appropriate when there is a low to medium flow rate over a wide range of product routings with varying distances. Since there is no cost to reconfigure a forklift-based system (other than communication to the operator), forklifts are most appropriate for variable product routings.

The answer is C.

46. Layout (A) has several long, high intensity flows that could be improved with a better facility layout. Layout (B) has backtracking flow between departments. Layout (C) has one long, high intensity flow.

Layout (D) has the best overall flow material as it has a central directed flow, its longer flows are all of a lower intensity, and there is no backtracking flow.

The answer is D.

47. Use the NIOSH formula to find the action limit.

$$\text{AL} = (90)\left(\frac{6}{H}\right)(1-(0.01)(|V-30|))$$
$$\times \left(0.7+\frac{3}{D}\right)\left(1-\frac{F}{F_{\max}}\right)$$

H is the horizontal distance of the hand from the body's center of gravity at the beginning of the lift, V is the vertical distance from the hands to the floor at the beginning of the lift, D is the distance that the object is lifted vertically, and F is the average number of lifts per minute.

The maximum permissible limit (MPL) in pounds per force is

$$H = 22 \text{ in}$$
$$V = 20 \text{ in}$$
$$D = 42 \text{ in} - 20 \text{ in} = 22 \text{ in}$$
$$F = \frac{1 \text{ lift}}{1.6 \text{ min}} = 0.625 \text{ lifts/min}$$
$$F_{\max} = 12 \text{ lifts/min}$$
$$\text{AL} = (90)\left(\frac{6}{22 \text{ in}}\right)(1-(0.01)(|20 \text{ in}-30|))$$
$$\times\left(0.7+\frac{3}{22 \text{ in}}\right)\left(1-\frac{0.625 \ \frac{\text{lifts}}{\text{min}}}{12 \ \frac{\text{lifts}}{\text{min}}}\right)$$
$$= 17.51$$
$$\text{MPL} = (3)(\text{AL}) = (3)(17.51)$$
$$= 52.53 \quad (53 \text{ lbf})$$

The answer is B.

48. To find the maximum height of the cart, use the anthropometric chart from the NCEES Handbook. Design the cart at the 5th percentile, since most people should be able to see over the cart. Use the eye height for women, since they are generally smaller than men.

$$\text{height of cart} = \text{average shoe height}$$
$$+ \text{5th percentile eye height women}$$
$$= 2 \text{ cm} + 138.3 \text{ cm}$$
$$= 140.3 \text{ cm} \quad (140 \text{ cm})$$

The answer is A.

49. The moment required by the body to lift the package can be found using the following equation.

$$\text{moment} = (\text{moment arm})(\text{weight})$$
$$\text{moment arm} = \frac{28 \text{ in}}{12 \ \frac{\text{in}}{\text{ft}}} = 2.333 \text{ ft}$$
$$\text{moment} = (2.333 \text{ ft})(40 \text{ lbf})$$
$$= 93.32 \text{ ft-lbf} \quad (95 \text{ ft-lbf})$$

The answer is A.

50. STE is the standard time for each element, R is the rating factor expressed as a percent, A is the allowances expressed as a percent, f is the frequency per cycle, STT is the sum of standard times for each element, and y is the average observed time for an element.

$$\text{STE} = \left(\frac{y}{f}\right)\left(\frac{R}{100\%}\right)\left(1+\frac{A}{100\%}\right)$$

For element 1,

$$\text{STE} = \left(\frac{0.14 \text{ min}}{1}\right)\left(\frac{110\%}{100\%}\right)\left(1 + \frac{20\%}{100\%}\right)$$
$$= 0.1848 \text{ min}$$

For element 2,

$$\text{STE} = \left(\frac{2.12 \text{ min}}{1}\right)\left(\frac{100\%}{100\%}\right)\left(1 + \frac{20\%}{100\%}\right)$$
$$= 2.544 \text{ min}$$

For element 3,

$$\text{STE} = \left(\frac{0.16 \text{ min}}{1}\right)\left(\frac{110\%}{100\%}\right)\left(1 + \frac{20\%}{100\%}\right)$$
$$= 0.2112 \text{ min}$$

For element 4,

$$\text{STE} = \left(\frac{5.28 \text{ min}}{5}\right)\left(\frac{110\%}{100\%}\right)\left(1 + \frac{20\%}{100\%}\right)$$
$$= 1.3939 \text{ min}$$

$$\text{STT} = 0.1848 \text{ min} + 2.544 \text{ min} + 0.2112 \text{ min}$$
$$+ 1.3939 \text{ min}$$
$$= 4.3339 \text{ min} \quad (4.34 \text{ min})$$

The answer is C.

51. n is the sample size, p is the proportion of observed time in an activity, $z_{\alpha/2}$ is the normal deviate required for a 2-tailed analysis with confidence of $1 - \alpha$, and D is the absolute error.

Using the following equation, examine each delay activity separately.

$$n = p(1 - p)\left(\frac{z_{\alpha/2}}{D}\right)^2$$

For the lack of material,

$$p = \frac{25}{200} = 0.125$$
$$D = 0.03$$

For a 2-tailed confidence of 95%, $\alpha = 5\%$, and $z_{\alpha/2} = 1.96$.

$$n = (0.125)(0.875)\left(\frac{1.96}{0.03}\right)^2$$
$$= 466.9 \quad (467)$$

For maintenance,

$$p = \frac{20}{200} = 0.10$$
$$D = 0.03$$
$$z_{\alpha/2} = 1.96$$
$$n = (0.10)(0.90)\left(\frac{1.96}{0.03}\right)^2$$
$$= 384.2 \quad (385)$$

For quality check,

$$p = \frac{5}{200} = 0.025$$
$$D = 0.03$$
$$z_{\alpha/2} = 1.96$$
$$n = (0.025)(0.975)\left(\frac{1.96}{0.03}\right)^2$$
$$= 104.04 \quad (105)$$

Select the largest n.

$$n = 467 \quad (470)$$

The answer is C.

52. For a 90% learning curve,

$$s = \frac{\ln(\text{learning rate})}{\ln 2} = \frac{\ln 0.9}{\ln 2}$$
$$= -0.152$$

K is a constant (for this problem, K is the 10 hr it takes to produce the first unit).

$$T_{\text{total}} = NT_{\text{ave}}$$
$$= \left(\frac{K}{1 + s}\right)\left((N + 0.5)^{1+s} - (0.5)^{1+s}\right)$$

The total time for completion of the 20 unit order is

$$\begin{array}{l}\text{total} \\ \text{time}\end{array} = \left(\frac{10}{1 - 0.152}\right)\left((20 + 0.5)^{1-0.152} - (0.5)^{1-0.152}\right)$$
$$= 146 \text{ hr} \quad (150 \text{ hr})$$

The answer is C.

53. Both profit/sales and sales/employee provide an indication of marginal profit per unit. The capacity used/max capacity provides the utilization of a resource, and profit/total investment indicates the return on investment derived from a company's activities. Each of the other answers contains at least one ratio that does not make sense (i.e., inventory/advertising cost, employees/department, and orders/delivery all provide no real useful relationship).

The answer is A.

54. The process is considered capable of meeting the designer's tolerances if the process's natural tolerance limits, $6\sigma_x$, are equal to or less than the spread between the tolerance limits; that is, $6\sigma_x$ (natural tolerance limits) $\leq (USL_x - LSL_x)$ (designer tolerance limits).

The answer is B.

55. If $6\sigma_x = USL_x - LSL_x$ and the process mean is set at μ_x halfway between the tolerance limits USL_x and LSL_x, only 3 parts in 1000 will fall outside of the specification limits as long as σ_x and μ_x do not shift.

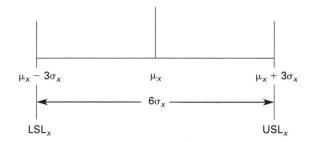

The answer is A.

56. The standard deviation of the difference between two normal random variates is $\sqrt{\sigma_B^2 + \sigma_A^2}$.

The answer is C.

57. Using a going with the flow (chart) and check sheets will enable effective data collection and creation of detailed flowcharts.

The answer is B.

58. Identify the problem category causes (e.g., machines, materials, operators, employees) and note the effect of each in terms of work, downtime, dollars lost, or scrap, and so on. Approximately 80% of the losses will be due to about 20% of the causes. Therefore, an organization should concentrate on 20% of the problem causes to reduce its inefficiencies.

The answer is D.

59. The distribution of the number of defectives in a random sample is binomial. The probability of accepting the lot is

$$P\{\text{defects} \leq 1\} = \sum_{d=0}^{c} \frac{n!}{d!(n-d)!} p^d (1-p)^{n-d}$$

$$
\begin{aligned}
c &= \text{total number of defects allowed} = 1 \\
n &= \text{random sample size} = 30 \\
p &= \text{fraction of defective items in a lot} = 3\%, \\
 &\quad \text{or } 0.03 \\
d &= \text{number of defects (i.e., in this case 0 and 1)}
\end{aligned}
$$

$$
\begin{aligned}
P\{\text{defects} \leq 1\} &= \frac{30!}{0!(30-0)!}(0.03)^0(1-0.03)^{30-0} \\
&\quad + \frac{30!}{0!(30-1)!}(0.03)^1(1-0.03)^{30-1} \\
&= 0.773
\end{aligned}
$$

The answer is B.

60. The 6σ spread for this process is based on the range data, R_i, using the following equation to estimate for σ.

$$
\begin{aligned}
\sigma &= \frac{\overline{R}}{d_2} \\
\overline{R} &= 4.60 \\
d_2 &= 2.059 \quad \left[\begin{array}{l}\text{found in the statistical}\\\text{quality control tables}\end{array}\right] \\
6\sigma &= (6)\frac{\overline{R}}{d_2} = (6)\left(\frac{4.6}{2.059}\right) \\
&= 13.4
\end{aligned}
$$

The answer is C.